Conserving Asia's Wildlife Treasure:

The Pheasants

Compiled by

Anita Chauhan

Edited by Dr. M. Shah Hussain

2014

PARTRIDGE

A Penguin Random House Company

To order additional copies of this book, contact
Partridge India
000 800 10062 62
www.partridgepublishing.com/india
orders.india@partridgepublishing.com

Contents

Acknowledgements

Wildlife scientists spend many years researching the ecology of birds and animals, providing clues for their effective conservation. Governments and institutions distill the research findings, apply them to the larger scenario of sustainable resource use, and develop policies and plans for action in their respective countries. It is important that this information becomes available in a simpler form to the people – to village- and urban-communities – so that they can participate in the process of wildlife and natural resource conservation. Thus, the field of conservation awareness and education holds an important place in the wildlife conservation conundrum. Books, posters, newsletters, events, public-space installations, exhibitions, websites, radio, television, etc, are the medium of communication with the masses.

Through this book, I am bringing to you information about wildlife species that are much valued in local or national cultures; and describing how the process of urbanisation and commercialisation has impacted wildlife. In our effort to strike a balance between resource use and conservation, ignorance and lack of empathy are a major hurdle. Therefore, conservation awareness has to become a popular revolution, reaching all strata of society, so that it supports sustainable resource-use through sustainable cities, sustainable forestry, sustainable agriculture, and eco-tourism, providing a better, long-term support to national economies.

I am grateful to all the people who have provided a solid backing to this project by encouraging me, and helping me prepare this book. Mr. Samar Singh (retired IAS officer, author, and formerly the President of World Pheasant Association-India) has reviewed the draft and provided suggestions in the early stages of its preparation. Dr. Rahul Kaul (Chief Ecologist, Wildlife Trust of India, Noida, U. P., India) has been supportive of the project despite his busy schedule. A discussion with Dr. M. L. Thakur (DST Scientist, Department of Biosciences, Himachal Pradesh University, Shimla, H. P., India) has helped me in presenting the information more clearly. Mr. Satpal Dhiman (HPFS Nodal Officer, Pheasant Conservation Breeding Programmes, Himachal Pradesh Forest Department, Shimla) has been very helpful by encouraging this project, and providing insights into pheasant conservation breeding activities in Himachal Pradesh, and helping me travel to Sarahan Pheasantry in Himachal Pradesh. Several other people in the H. P. Forest Department have been helpful in my understanding of wildlife conservation in the state. I thank them all for their cooperation. Dr. V. K. Santvan (Assistant Professor, Environmental Science, Institute of Integrated Himalayan Studies, H. P. University, Shimla) has been supportive of the book project. Dr. M. Shah Hussain (Scientist In-charge, Aravalli Biodiversity Project, University of Delhi, and Hon. Gen. Secretary, WPA-India), a pheasant expert who has spent many years studying the ecology of pheasants in the Uttarakhand Himalayas, has very kindly reviewed the draft. He has also edited the final draft of this book. I thank Dr. Shah Hussain and his wife, Dr. Aisha Sultana for their support. Dr. Faiyaz A. Khudsar (Scientist Incharge, Yamuna Biodiversity Project, University of Delhi, New Delhi; and senior coursemate from my student-days) has been enthusiastic and cooperative about this book project.

The Maharaja Fatesinghrao Gaekwad Library and Documentation Centre at WWF, New Delhi has been extremely useful, and I thank their helpful staff members. I also appreciate the cooperation of the State Museum, Shimla, Himachal Pradesh, by letting me view the photos of traditional artwork featuring the wildlife of the state from their museum collection. Some of these photos can be seen on my Flickr.com blog @ AnitaWPANewDelhi. I also acknowledge the help provided by the various state emporia at Baba Kharag Singh Marg, Connaught Place, New Delhi, by letting me photograph the handicrafts featuring Indian pheasants. I am thankful to the members of the Phasianidae Group on Flickr.com, especially to Staven, and to others who have contributed photos.

I also acknowledge the prompt permissions provided by all the individuals, institutions, and publishers that were contacted for the quoted text, and I am profoundly grateful to them. Thanks also to my contacts at Partridge Publishing, India (Mark Montez, Nelson Cortez, and Gemma Ramos) for their help.

Lastly, I thank my parents, Smt. Pramila and Shri D. S. Chauhan, for their support for this book, and my relatives in Shimla, Rampur, and Kotgarh who have helped me explore some of the pheasant habitats in Himachal Pradesh, India while researching for this book.

Preface

In the past few decades, there has been an increase in interest in bird-watching in India and other countries. People of all backgrounds – IT professionals, teachers, students, doctors, housewives – are donning their outdoors gear on the weekends to enjoy the natural environment, and to educate themselves about the flora and fauna in their leisure time.

As all of you would agree, conservation awareness among the masses is an essential component of the biodiversity conservation programme of a country. This book is especially written for making the city-dwelling people and the student community aware about the habits and the conservation of an important component of our biodiversity – the pheasants – a group of ground-dwelling birds of the Family Phasianidae (which also contains other groups of game-birds such as partridges, turkeys, guans, and megapodes). While the earlier books on this subject have been written as a biologist's, an aviculturist's, or a game-hunter's reference, this book is for the everyday naturalist.

The mention of 'pheasants' invariably brings to people's minds words such as 'grandeur' and 'elusive'. Indeed, the pheasants have these qualities. These magnificent birds have been a part of the human culture for centuries, and yet, are now facing threats to their survival. The universal problems of shrinking wildlife habitat, over-hunting, and isolation of populations, are also taking their toll on the pheasants. As the readers will learn in the following chapters, the pheasants are primarily distributed in Asia, and their forest-habitat is under pressure from the growing human population. Therefore, protection of their natural habitat is the first step in saving these species from extinction. Several international and national agencies are involved in conservation activities; and learning about these activities by the readers will eventually help to further strengthen them.

An interesting facet has been added to the chapters in the form of excerpts from the early 20th century American naturalist Dr. William Beebe's wonderful book *Pheasant Jungles.* Dr. Beebe wrote an account of his journeys into the forests in Asia in search of the pheasants (while researching for his monograph on pheasants). The excerpts have been woven-in to make the reading experience more vivid, and to help the readers appreciate the following scientific information better.

The last chapter in this book describes the conservation projects that are being implemented by the respective State Forest Departments in India. The information provided will, hopefully, also help to kindle wanderlust in readers, and help them plan their journeys to zoos and pheasant habitats, in order to familiarize with these species.

I know that students and wildlife enthusiasts of all ages will find this book interesting. In the UN 'Decade of Education for Sustainable Development (2005-2014)', I also hope that the book will set a trend for more such conservation-education literature on endangered fauna.

Anita Chauhan, March 2014, New Delhi

Email: pheasantsbookrespons@gmail.com

Chapter 1

Discover the Pheasants

The plumage of this tragopan is beautiful beyond most birds – orange and gold, brown and black, with a score of constellations of silver stars scattered over all, from neck to tail.

-   From *Pheasant Jungles* by William Beebe

Fig. 1 Detail from a painting of western tragopan (*Tragopan melanocephalus*) male (on the rt.) and female, painted by Archibald Thorburn, in William Beebe's *A Monograph of the Pheasants* (1918-22).

Chapter 1

Discover the Pheasants

Most of us are aware that the elegant blue peafowl - also called 'peacock' - is India's National Bird. But many of us do not know that the peacock belongs to a group of birds known as the 'pheasants', and that there are about 50 different species in this group. These forest-dwelling birds carry out the ecological function of seed dispersal, control of insects and small reptiles, and are a source of food for a variety of predators.

What are Pheasants?

Pheasants are moderately large birds, comparable in size to domestic chickens. They primarily feed on the ground, and eat plant matter, insects, etc. In most of the species of pheasants, the males have colorful, iridescent plumage. Pheasants build simple nests on the ground with leaf- and stem-scraps, and a few species also construct nests on trees or bushes. The pheasants are generally non-migratory; and they are either monogamous or polygamous (Johnsgard, 1999). Their courtship displays are perhaps the most spectacular and alluring among all the birds in the world. This makes them prime subjects of depiction in art and literature in their indigenous range countries. Some examples of these remarkable species are the blue peafowl, the green peafowl, the brown eared pheasants, the crested fireback pheasant, and the silver pheasant.

Where do the Pheasants Live?

Most of the species of pheasants are found in Asia - in the montane temperate forests (of Indian Himalayas, mountains of Pakistan, Nepal, Bhutan, China, Japan, and Southeast Asia), temperate grasslands (of central Asia), and tropical forests (of India, Sri Lanka, and Southeast Asia); and one of the species, the Congo peacock *Afropavo congensis*, occurs in the tropical forests of Congo in Africa (see Appendix 7). Pheasant habitats are still some of the least-studied forest ecosystems in the world - and yet - they are one of the bird groups most seriously affected by habitat destruction and over-hunting (Johnsgard, 1999). As we will read in the following chapters, almost 2/3rd of the entire group of pheasant species are listed as 'Endangered', 'Vulnerable', or 'Near-threatened' in the IUCN Red List (see Appendix 1).

COMMON PHEASANT (*Phasianus Colchicus*).

Fig. 1.1 A sketch of the ring-necked pheasants by T.W. Wood, in 'Pheasants: Their Natural History and Practical Management' 2nd edition, 1881, by W.B. Tegetmeier.

Yet, the pheasants have been of great benefit to humans; and the most commonly domesticated species of bird in the world - the domestic fowl - has originated from a pheasant species called the red junglefowl *Gallus gallus*. Other pheasant species - such as the ring-necked pheasant *Phasianus colchicus* (a.k.a. common pheasant) have been introduced from Asia into the USA and Europe, where they are either kept as ornamental pets by aviculturists, or bred as game birds (Johnsgard, 1999). Millions of ring-necked pheasants are then released into hunting reserves every year for the sport of hunting. Other, rarer pheasant species have been kept in private aviaries and government-owned zoos for more than a century in Europe, in USA, and in Australia. Information about exotic pheasants kept in captivity is recorded in studbooks to breed them for the purpose of cage-bird trade, and also for *ex-situ* conservation programs. This is described further in the following chapters.

Classification of Pheasants

The vertebrate animal kingdom is grouped into classes, such as - class Amphibia (consisting of frogs, toads and salamanders), class Reptilia (consisting of snakes, lizards and crocodiles), class Aves (consisting of birds), and class Mammalia (consisting of mammals). The animals in each class are further classified into orders, families, genera (singular, genus), and species.

The pheasants are classified into class Aves, order Galliformes, and family Phasianidae. The scientific nomenclature of animals follows a Binomial System (consisting of Latinized names) in which the 1st of an animal's two or three names is that of the genus, and the 2nd is that of the species, while a 3rd is sometimes used for sub-species (Grewal, 1993). Each genus may contain a number of species. For example, the scientific name of the Ceylon junglefowl (belonging to class Aves, order Galliformes, family Phasianidae, and genus *Gallus*) is *Gallus lafayettei*, and that of the grey junglefowl is *Gallus sonnerati*, 2 of the 4 species in this genus.

The order Galliformes contains 286 species of birds that are known as the 'game-birds', as most of these large-sized birds have historically been hunted by humans for their eggs and meat. The word 'galliform' is derived from the Latin term *gallina*, which means 'hen'. The 7 families in this order are –

Order Galliformes

Sub order Craci

Family Megapodiidae – Megapodes

Family Cracidae – Chachalaca, guans, curassows

Sub order Phasiani

Family Meleagrididae – Turkeys

Family Tetraonidae – Grouse

Family Odontophoridae – New World Quails

Family Phasianidae –

    Subfamily Perdicinae - Partridges, quails, spurfowl

    Subfamily Phasianinae – Pheasants, peafowl [16 genera, 49 species]

Family Numididae – Guineafowl (Madge and McGowan, 2002)

The physical characteristics that differentiate the pheasants from other galliform birds, such as grouse and quail, are – pheasants are generally larger, exhibit greater sexual dimorphism, and the males usually have iridescent feathers and highly graduated tails. The tarsi (the part of the leg of a bird below the thigh) and nostrils in pheasants are unfeathered. Unlike spurfowl, only the male pheasants have spurs on their legs, and female pheasants lack spurs. Pheasants have a smooth and unserrated lower mandible (lower beak); and the males lack inflatable air sacs on the neck that are typical of male grouse. The molting pattern of the feathers of wings and tail in pheasants is also different from other galliform birds (Johnsgard, 1999).

Zoogeographical Distribution of Pheasants

Most of the species of pheasants are found in Asia - in the Indo-Malayan zoogeographic realm and the south-eastern part of the Palearctic realm. Excepting the single species of Congo peacock *Afropavo congensis* found in the Afrotropical realm, all pheasant species occur in Asia - between the eastern coast of Black Sea in the west and Japan in the east - and extending northward to Mongolia and southward to the Lesser Sunda Islands (Johnsgard, 1999).

The pheasants occur in areas that receive more than 60 inches of annual precipitation in temperate and tropical settings, and those that are at least partially wooded with montane or lowland forest (Johnsgard, 1999). A patchwork quilt of numerous species-rich forests of different types covers the landscape, in which the birds and animals, including our spectacular yet secretive pheasants, form characteristic assemblages. A concise description of the realms now follows, which will help us to understand the characteristics of the habitats in which the various species of pheasants occur.

The Indo-Malayan Realm

The Indo-Malayan realm's natural boundaries contain tropical Asian countries. The realm includes 3 main floristic regions: the Indian sub-continent (comprising Pakistan, India, Nepal, Bhutan, Bangladesh, and Sri Lanka); Indo-China (comprising Myanmar, Vietnam, Laos, Cambodia, Thailand, and the tropical southern fringe of China with Taiwan); and Malesia (comprising Malaysia, Philippines, Singapore, Brunei, and Indonesia) (Watkins et al., 1997).

The physical factors that determine climate are altitude, latitude, and rainfall. The climate of the realm is very diverse: from the dry Thar Desert at one extreme, to the torrential rainfall region of the Assam hills and the lush tropical rainforests of Borneo at the other. The realm also exhibits a wide range of altitude – from sea level to the Himalayas (Watkins et al., 1997).

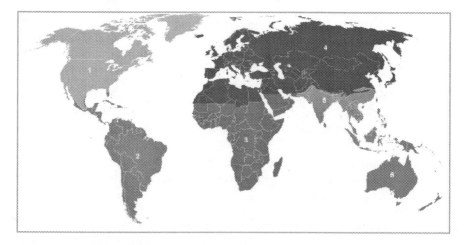

Legend: 1. Nearctic ecozone, 2. Neotropic ecozone, 3. Afrotropic ecozone, 4. Palearctic ecozone, 5. Indomalayan ecozone, 6. Australasian ecozone (Oceania and Antarctic zones are not shown).

Fig. 1.2 A map of the 8 zoogeographical zones of the world.

Fig. 1.3 A map showing the location of Democratic Republic of Congo in central Africa.

Table 1.1 Different types of forests are found in the 3 zoogeographic realms where the pheasants occur.

Realm	Region	Forest Type	Sub-Type	Examples of Pheasant Species
Indo-Malayan Realm	Indian Subcontinent [Pakistan, India, Nepal, Bhutan, Bangladesh, (Sri Lanka, and Western Ghats in India containing Tropical Rainforest)]	Himalayan Forest	Wet Tropical Foothill Forest	Blue peafowl, red junglefowl
			Mixed and Alpine Forest	Western tragopan, Satyr tragopan, cheer, koklass, Himalayan monal, kalij, red junglefowl, Sclater's monal, Blyth's tragopan, Temminck's tragopan, grey peacock pheasant, grey eared pheasant, Mrs. Hume's pheasant

7

	Indo-China (Myanmar, Vietnam, Laos, Cambodia, Thailand, parts of China)	Southeastern Indo-China Dry Evergreen Forest		Germain's peacock-pheasant
		Northern Indo-China Moist Forest		Blyth's tragopan, Temminck's tragopan, blood pheasant, Siamese fireback, white eared pheasant, green peafowl
		Tropical and Subtropical Moist Broadleaf Forest		Cabot's tragopan, silver pheasant, Swinhoe's pheasant, Mikado pheasant
	Malesia (Malaysia, Philippines, Singapore, Brunei, Indonesia)	Tropical Rainforest	Lowland Forest	Bornean peacock pheasant, great argus pheasant, Palawan peacock pheasant, Bulwer's pheasant, grey junglefowl, green junglefowl, crested fireback and crestless fireback pheasant, crested argus pheasant, mountain peacock pheasant
			Montane Forest	
		Deciduous Forest		
Palearctic Realm	China, Japan, Mongolia, Korea, Russia	Temperate Forest	Coniferous Forest	Lady Amherst's pheasant, Reeve's pheasant, golden pheasant, copper pheasant, Temminck's tragopan, common pheasant
			Broadleaf and Mixed Forest	
Afrotropic Realm	Congo DRC	Tropical Rainforest		Congo peacock

Indian Sub-continent

The Indian Sub-continent exhibits a variety of climates, landscapes, and forests. The Himalayan mountain ranges, the deciduous forests of alluvial plains, the Thar Desert, the rainforests of Western Ghats, the dry Deccan plateau, and the Mangroves are the major landforms. Out of these, the Himalaya and the Western Ghats are important pheasant habitats.

A. The Himalayas form a continuous chain of tall mountain ranges that stretch over 2400 kilometers - from the Indus River in the west - to the Brahmaputra River in the east. 'Himalaya' is a Sanskrit term which means 'abode of snow', and this mountain chain has 8 of the world's highest mountains with a height of over 8000 meters. The mountains affect the climate of the sub-continent by deflecting the moisture bearing clouds, which results in precipitation over much of the region; and this precipitation creates the great rivers of Asia: the Indus, the Ganges, the Tsangpo/Brahmaputra, the Salween, the Mekong, and the Yangtze. Along with their tributaries, these rivers water the landscape that sustains millions of people in the mountains and in the floodplains. The rivers are utilized for hydropower, irrigation, supplying drinking water, and for supporting industrial units along their lengths (Zurick et al., 2005).

Fig 1.4 A photo of the Himalayan Mountains in the state of Himachal Pradesh in northern India, by Anita Chauhan, 2011.

According to the theory of continental drift, about 130 million years ago, the Gondwana continent started drifting toward the main Asian continental plate, resulting in a collision (about 60 million years ago) and the lifting of the oceanic crust of the intervening ancient Tethys Sea, creating the Himalayas. Evidence of this event can be found in the landscape – strata of exposed rocks showing buckling and folding, fossilized brachiopods, corals and skeletal fish trapped at the foot of glaciers, and seashells and ammonites. So, the mountain ranges of the Himalayas are now the outer skin, so to speak, of the South Asian geological plate that continues to submerge below the Asian continental plate, and the Himalayas continue to rise about 2 cm per year (Zurick et al., 2005).

There is no single climate in the Himalayas; climate becomes kaleidoscopic with the changing altitude and orientation to the sun. From north to south, the mountains are spread across temperate to subtropical zones. Apart from the altitude and latitude, the other major determinant of climate in the Himalayas is their location relative to the Asian monsoon airflow. In the western Himalayas and in the adjoining Karakoram Range are found some of the world's largest glaciers, and their presence is attributed to the local topography, with its large basins, as well as to the fact that the western Himalayas get considerably more precipitation from winter storms, producing locally heavy snowfalls. The intensity of the winter storms tapers off toward the east, and is limited to the highest summits in eastern Nepal, Bhutan, and Arunachal Pradesh. In the eastern and central Himalayas, most of the precipitation falls in the summer, and it reduces after 2000 meters elevation (Zurick et al., 2005).

There is a vertical zonation of vegetation types in the Himalaya, and there are also horizontal differences across the length of the range, spanning alpine to wet tropical. The tropical deciduous sal tree *Shorea robusta* is a valuable lowland timber species that is widespread in this range. Tropical broadleaf forests also include oak *Quercus leucotrichophora*. Subtropical pines such as Chir pine *Pinus roxburghii* occur at intermediate elevations in the middle mountains, giving way, at elevations above 2000 meters, to moist mixed temperate forests. Mixed temperate forests in the western part of the range include trees, such as deodar *Cedrus deodara*, Himalayan horse chestnut *Aesculus indica*, maples *Acer sp.*, oaks *Quercus sp.*, rhododendrons, and economically important species, such as Himalayan blue pine *Pinus wallichiana* and Himalayan cypress *Cupressus torulosa*. In the central and eastern sectors of the Himalayas, the moist subtropical and temperate forests contain numerous species of bamboo; and a variety of pines, spruce *Picea*, fir *Abies*, and juniper *Juniperus* occur in the upper temperate and subalpine zones, which converge with the alpine level in a mix of dwarf conifers, willows *Salix*, and birches *Betula* (Zurick et al., 2005).

Timber extraction on a large scale during 1890-1945 by the British colonists, for building railroads in the plains, and for building summer resort towns, such as Shimla and Nainital, was responsible for denudation of forests in Himachal Pradesh, and Garhwal and Kumaon regions of Uttarakhand in western Himalayas. In recent years though, unlike in the Southeast Asian region, where hardwood timber export has caused the denudation of forests, the Himalayan region has experienced loss of forest cover due to an increase in human population that is dependent on the forests for firewood, fodder, and construction wood, and due to the denudation of forest land for agricultural and urban expansion. 'The forests have been declining at the rate of 1% per year across the Himalayas since the 1960s' (Zurick et al., 2005).

Global warming is also a threat to the Himalayan ecosystem. As precipitation patterns change, the mixed temperate and alpine forests, and perennial streams, which are dependent on water trickling down from melting snow in the higher reaches will be affected if the snow melts too quickly, or if there is little snow.

The Himalayan mountains are home to many endemic and other wildlife, such as – blue sheep, snow leopard, ibex, musk deer, urial, red panda, wolves, black bear, leopard, langur, wild boar, barking deer (Zurick et al., 2005), and pheasant species- e.g. western tragopan *Tragopan melanocephalus*, cheer pheasant *Catreus wallichii*, koklass pheasant *Pucrasia macrolopha*, Himalayan monal *Lophophorus impejanus*, kalij pheasant *Lophura leucomelanos*, red junglefowl *Gallus gallus*, Sclater's monal *Lophophorus sclateri*, Blyth's tragopan *Tragopan blythii*, grey peacock pheasant *Polyplectron bicalcaratum*, and Mrs. Hume's pheasant *Syrmaticus humiae*. The southern parts of the range, i.e. the subtropical foothills, provide a habitat for a rich assemblage of wildlife that includes Asiatic elephant, one-horned rhino, gaur, sloth bear, tiger, and various reptiles, birds, and fishes (Zurick et al., 2005).

B. Tropical rain forests are found on the mountains of Western Ghats, Andaman and Nicobar Islands of India, and south west Sri Lanka. These forests contain many endemic species of plants and animals, and are a source of wild-genotypes of important food plants and other economically important plants. The grey junglefowl *Gallus sonneratii* is found in the Indian peninsula. They are found in thickets, on the forest floor, and in open scrub. The Sri Lanka junglefowl *Gallus lafayetii* occurs in Sri Lanka - in forest and scrub habitats. The Nicobar groups of islands are home to the Nicobar megapode - no pheasants are found on these islands.

The Western Ghats is a mountain range along the western side of Indian peninsula. It is a UNESCO World Heritage Site, and is one of the 'hotspots' of biological diversity in the world. The range runs north to south along the western edge of the Deccan Plateau. The range starts near the border of Gujarat and Maharashtra south of the Tapti River, and runs approximately 1,600 km through the states of Maharashtra, Goa, Karnataka, Tamil Nadu, and Kerala, ending at Kanyakumari at the southern tip of India. The Western Ghats are not true mountains, but are the faulted edge of the Deccan Plateau. After the break-up of the super continent of Gondwana (some 150 million years ago), the South Asian continental plate broke away from Madagascar, and the west coast of India was formed (somewhere around 100 to 80 million years ago). The western coast of India would have appeared then as an abrupt cliff some 1,000 m in elevation. The highest point in Western Ghats is about 2700 meters in elevation (Wikipedia).

A total of 39 properties including national parks, wildlife sanctuaries, and reserve forests are designated as world heritage sites in the Western Ghats - 20 in Kerala, 10 in Karnataka, 5 in Tamil Nadu, and 4 in Maharashtra. The area is one of the world's top ten 'biodiversity hotspots', and has over 5000 species of flowering plants, 139 mammal species, 508 bird species, and 179 amphibian species; and it is likely that many undiscovered species live in the Western Ghats. At least 325 globally threatened species occur in the Western Ghats. Endangered mammals of the Western Ghats include - Malabar large-spotted civet, lion-tailed macaque, gaur, tiger, Asian elephant, and Nilgiri langur. Tree species found in the lowland forests include family Dipterocarpaceae, while trees of family Lauraceae are found in the highlands (Wikipedia).

Loss of forest land due to plantations of tea, coffee, ginger, cardamom, etc., and rapid urbanisation is a cause of concern in the Western Ghats. The Sri Lankan forests are similarly rich in species of endemic fauna - amphibians, reptiles, birds, mammals - as well as endemic flora, though their forest habitat has been fragmented due to human landuse (Wikipedia).

World Heritage Sites in the Indian subcontinent are - Nanda Devi National Park and Biosphere Reserve and Valley of Flowers National Park in western Himalayas in India, Royal Chitwan National Park in Nepal, Sinharaja Forest Reserve and Central Highlands in Sri Lanka, Sundarbans in Bangladesh, Western Ghats in South India, and Kaziranga National Park and Manas Sanctuary in northeast India. Great Himalayan National Park (in Himachal Pradesh, India) has also been nominated as a World Heritage Site.

Indo-China

3 main forest types are found in this region (WWF, 2013 a).

A. The Southeastern Indochina Dry Evergreen Forests region (Monsoon forest) occurs in a broad band across northern and central Thailand into Laos, Cambodia, and Vietnam. Dry evergreen forest is more appropriately called semi-evergreen forest because a significant proportion of canopy tree species are deciduous at the height of the dry season. Extensive areas of this ecoregion in Laos and Cambodia occur with a largely deciduous forest canopy. This type of forest occurs in climatic regions where the mean annual rainfall is generally between 1,200 and 2,000 millimeters, and a significant dry period of 3-6 months occurs each year.

Forests of this region have a characteristic tall and multilayered forest structure (similar to that of lowland evergreen rain forest of Malesia). In comparison to evergreen rain forests though, the species diversity is lower, and the canopies and understory more open. Although dry evergreen forests are floristically related (have many plant families and genera in common) to lowland evergreen rain forest communities (of

Malesia), they exhibit a unique floristic structure that is rich in species that are endemic to mainland Southeast Asia.

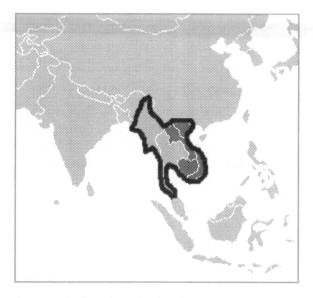

Fig. 1.5 Map showing the location of Indo-China region in Asia (Myanmar, Vietnam, Laos, Cambodia, Thailand, and parts of China) outlined in black.

The canopy of semi-evergreen forests generally is multilayered, and reaches a height of about 30-40 meters, with an open structure. Dipterocarp trees are a major component of the forest structure, and form emergent tree canopies, with such species as *Dipterocarpus alatus, D. costatus, Hopea odorata, Shorea guiso, S. hypochra*, and *Anisoptera costata*. Many dipterocarp species are now threatened with extinction due to over-exploitation. Other giant emergents are species of *Ficus, Tetrameles nudiflora*, and *Heritiera javanica*, each forming large buttresses at the base of their trunks. In broad valley areas, semi-evergreen forest often occurs as a fringing gallery forest along streams, grading out into deciduous dipterocarp forest on drier sites with shallower soils. Unusually large trees such as *Dipterocarpus turbinatus, D. alatus*, and *D. costatus* once formed dense stands in these forests, with a variety of other dipterocarps in ravines in such habitats. These species begin to drop out at about 700 meters elevation, and are replaced by montane species.

Bamboos are common in the forest, particularly as colonizers of open gaps after disturbance. Palms are present, most notably along watercourses, but less abundant and diverse than in lowland tropical rain forest habitats. Lianas are abundant, and the understory is less complex than in lowland rain forest, and species richness is also lower.

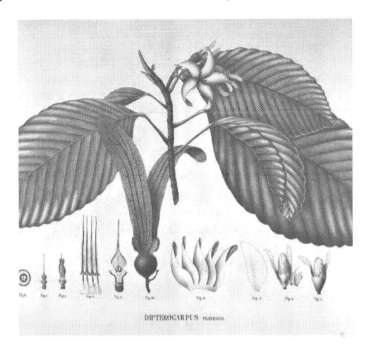

DIPTEROCARPUS ᴛᴇᴀᴍᴇᴀᴠᴇ.

Fig 1.6 A drawing of two-winged fruit, leaves, and flower of a *Dipterocarpus* sp. tree from *Flora Javae* by C.L. Blume and J. B. Fischer.

The Southeastern Indochina Forests region is globally outstanding for the large vertebrate fauna it harbors within large intact landscapes. Among the impressive large vertebrates are the Indo-Pacific region's largest herbivore, the Asian elephant, and largest carnivore, the tiger. The species list includes the second known population of the critically endangered Javan rhinoceros (comprising a handful of animals in Vietnam's Cat Loc reserve), and Eld's deer, banteng, gaur, clouded leopard, douc langur, pileated gibbon, southern serow, orange-necked partridge, Germain's peacock pheasant *Polyplectron germaini*, common leopard, and Malayan sun bear.

B. The Northern Indo-China Moist Forests (Temperate Deciduous forest) which include the Yunnan Plateau subtropical evergreen forests (southeastern China), and Northern Indo-China subtropical forests (China, Laos, Myanmar, Thailand, and Vietnam), are also rich in species of pheasants (white-eared pheasants *Crossoptilon crossoptilon* and green peafowl *Pavo muticus* are found in the former, and Blyth's tragopan *Tragopan blythii*, Temminck's tragopan *Tragopan temminckii*, ring-necked pheasant *Phasianus colchicus*, blood pheasant *Ithaginis cruentus*, and Siamese fireback *Lophura diardi* in the latter region), and mammals such as Asiatic black bears, black gibbons, bamboo rats in the former, and Tonkin snub-nosed monkey, tiger, Asian elephant, and gaur occur in the latter region.

The Yunnan Plateau has several types of oak and laurel trees. The leaves of these trees are hairy, which helps the trees to conserve water and remain green during the long dry season. In Northern Indochina, open-canopy pine forests in the higher elevations give way to patches of tropical forests in the moist valleys below. Tropical seasonal rain forests sustain dipterocarp and fig species, while tropical montane forests are comprised of *Alstonia scholaris*, *Phoebe puwenensis*, and *Litsea pierrei* trees. Oak, tea, and chestnut species occur in the evergreen broad-leaved woodlands.

C. Other types of forests in this region are - Tropical and Subtropical Moist Broadleaf Forests (including Taiwan subtropical evergreen forests and South Taiwan monsoon rain forests, Jian Nan subtropical evergreen forests, South China-Vietnam subtropical evergreen forests, and Hainan Island monsoon rain forests). The flora and fauna includes Ginkgo tree, dawn redwood, cycads, and rhododendrons, Cabot's tragopan *Tragopan caboti*, silver pheasant *Lophura nycthemera*, Swinhoe's pheasant *Lophura swinhoii*, Mikado pheasant *Syrmaticus mikado*, Taiwanese macaque, serow, Sitca deer, and Asiatic black bear.

Most of the forests in Vietnam have already been replaced by plantations. Shifting agriculture has further degraded some areas of this ecoregion. But the greatest threats were from large-scale logging concessions (now revoked) that had been granted to multinational companies by the Cambodian government. Hunting, to supply the huge wildlife trade, had created empty forests throughout most of the region. The use of small homemade crossbows to kill small mammals for local consumption, and bombs (hidden in baited traps to kill tigers) and pitfall traps (for elephants), has taken a heavy toll on wildlife. War and conflict in this region caused the easy availability of automatic weapons, which replaced the crossbows. Read more details in Chapter 4.

UNESCO World Heritage Sites in the Indo-China region include Thungyai-Huai Kha Khaeng Wildlife Sanctuaries and Dong Phayayen-Khao Yai Forest complex in Thailand, and Phong Nha-ke Bang National Park in Vietnam.

Malesia

The countries in this ecoregion include: Malaysia, Philippines, Singapore, Brunei, and Indonesia. In this region, rainfall is important both in terms of annual total as well as seasonal distribution. Equatorial lowland areas that receive at least 6 cm of rain each month can support evergreen rainforest. Areas with marked dry seasons can support deciduous (monsoon or subtropical) forest where the total annual rainfall is very high. The tropical rain forests in Malesia are similar to the montane moist forest stretching along the foothills of the eastern Himalayas into the mountains of Indo-China, and occur in isolated patches on the highest equatorial mountains in the archipelago. These forests can be divided into lowland and montane. The distinction between the two lies at about 1000 meters altitude at the equator, and becomes progressively lower going towards greater latitudes (Watkins et al., 1997).

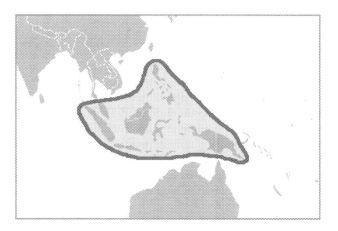

Fig. 1.7 Map showing the location of Malesia region in Asia.

The first two floristic regions have about 15,000 plant species each. The Malesian flora is conservatively estimated to contain 25,000 species of flowering plants, about 10% of the world's total. Floristically, Malesia is the richest and most important regions. Malesia contains one of the largest blocks of tropical rainforest in the world- second only to the Amazon region. About 40% of the genera found in Malesia, and even more of the species, are endemic to the region. The biggest plant family is Orchidaceae, with 3000-4000 species. Some 500 species of trees of the family Dipterocarpaceae are found in the region (Watkins et al., 1997).

Many food-crop plants, and other plants and animals of economic importance came originally from this realm. India, Burma, and Southeast Asia are rich centers of genetic variation in cultivated fruit trees and their wild relatives, only about 30% of which have been cultivated so far. Forest plants provide beverages, fibers, fruits, gums, resins, oils, bamboos, rattans, spices, and vegetables. Fruits such as banana, citrus, mango, mangosteen, and spices, such as cardamom, cinnamon, clove, ginger, nutmeg, pepper, turmeric, as well as fibers, such as ramie and jute, all originated in this region. The forests are rich in medicinal plants, and in ornamental plants, such as orchids and palms. The dipterocarp forests of Southeast Asia are the largest source of hardwood in international trade. In Peninsular Malaysia, about 4/5th of the timber produced comes from dipterocarps. The leading species from the monsoon forests is teak (Collins et al., 1991). We will read about the destruction caused due to over-exploitation of timber in Chapter 4.

Having derived its original fauna from the Palearctic, Africa, and to a lesser extent, from Australia and the drifting South Asian continental plate, the Malesia region has been a recent and dynamic center for evolution. A history of relatively stable equatorial conditions, combined with dynamic geological changes in which land connections have been repeatedly formed and broken, has allowed isolated populations to radiate. These events have led to the enormous species richness and the high levels of local endemism characteristic of the region. For example, Indonesia alone has more species of birds and trees than the whole continent of Africa (Watkins et al., 1997).

Fig. 1.8 Photo of the rainforest in Sumatra, Indonesia showing dense forest and emergent Dipterocarpus trees along a stream, by Stefan Ottomanski, 2010.

Fig. 1.9 Rhinoceros hornbill in the tropical rainforest of Malaysia, by Kevin Lin, 2011.

The forests of the Malesia ecoregion are rich in wild relatives of domesticated animals, many of which are threatened by over-hunting and loss of suitable habitat. Some of these wild animals are: Javan warty pig, serow, gaur or seladang, banteng, kouprey, sao la, wild water buffalo, anoa, and tamaraw. Of less direct value to mankind and his domesticated species, but of much concern in terms of Asia's natural heritage, are the carnivores (civets, tiger, clouded leopard, Bornean bay cat, marbled cat, flat-headed cat, rusty-spotted cat, and Asiatic golden cat), rhinoceroses, elephants, and primates (Collins et al., 1991), and pheasants such as Bornean peacock pheasant *Polyplectron schleiermacheri*, great argus pheasant *Argusianus argus*, crested argus pheasant *Rheinardia ocellata*, Palawan peacock pheasant *Polyplectron napoleonis*, Bulwer's pheasant *Lophura bulweri*, and green junglefowl *Gallus varius*.

The countries of this region have numerous threatened birds, most of them residents of the lowland rain and monsoon forests. Particularly notable are Vietnam (with 34 threatened species), Thailand (39), Malaysia (34), Indonesia (126), and Papua New Guinea (24). Some species are now extremely rare, notably the white-eyed river martin from Thailand, Vo Quy's pheasant *Lophura hatinhensis*- known only from 2 specimens in Vietnam, and Gurney's pitta, recently rediscovered in a 1.6 sq. km Thai forest after half a century. The enigmatic double-banded argus pheasant *Argusianus bipunctatus*, described only from a feather collected in the 19th century from Java (and now displayed in the London Natural History Museum, England), is presumed to be extinct as it has not been sighted in the region since (Collins et al., 1991).

The World Heritage Sites in Malesia ecoregion include: Komodo National Park, Lorentz National Park, and Ujung Kulon National Park in Indonesia, and Kinabalu National Park and Gunung Mulu National Park in Malaysia. Kinabalu National Park (or Taman Negara Kinabalu in Malay), established as one of the first national parks of Malaysia in 1964, is Malaysia's first World Heritage Site designated by UNESCO in December, 2000 for its "outstanding universal values" and its role as one of the most important biological sites in the world. Located on the west coast of Sabah, on the island of Borneo, it covers an area of 754 square kilometers surrounding Mount Kinabalu, which at 4,095.2 meters, is the highest mountain on the island of Borneo. The park is one of the most popular tourist spots in Malaysia. In 2004, more than 415,000 people visited the Park. Gunung Mulu National Park in Sarawak, Malaysian Borneo, is the 2nd UNESCO World Heritage Site. It includes enormous caves and karst formations in a mountainous equatorial rainforest. The national park is named after Mount Mulu, the second highest mountain in Sarawak.

Interestingly, the Wallace's line (lying between Borneo and Sulawesi (earlier called Celebes)) rather effectively separates the distribution patterns of the pheasants and megapode species, the latter extending from the coastal islands of Borneo to throughout much of the Australian region (Johnsgard, 1999).

The Palearctic Realm

The Palearctic realm includes mostly boreal and temperate climate ecoregions, which run across Eurasia from Western Europe to the Bering Sea. Forests in China, that form the southeastern border of this realm with the Indo-Malayan realm, are home to many species such as Lady Amherst's pheasant *Chrysolophus amherstiae*, Reeves's pheasant *Syrmaticus reevesii*, and golden pheasant *Chrysolophus pictus*. These forests are of 2 types (WWF, 2013 a) -

A. Temperate Coniferous Forests (including Hengduan Mountains subalpine conifer forests, Qionglai-Minshan conifer forests, and Nujiang Langcang Gorge alpine conifer and mixed forests). Among the rare plant species found here is the dawn redwood (*Metasequoia glyptostroboides*), which until recently was presumed extinct. Other tree species include the Chinese yew (*Taxus chinensis*), Manglietta (*Manglietia fordiana*), and Chinese cedar (*Cryptomeria fortunei*). Many of these plants represent the last vestiges of once widespread vegetation types. The fauna of this region includes Lady Amherst's pheasant *Chrysolophus amherstiae*, giant panda, red panda, and spotted linsang.

Copper pheasant *Syrmaticus soemmerringii* is found in Japan, in the wooded hills near Mt. Fuji. The vegetation here is mainly *Cryptomeria japonica*, pines, bamboo, maples, and oaks. During the cold winters it decends to lower elevation where it is found with the Japanese green pheasant *Phasianus c. versicolor* near agricultural fields (Beebe, 1922).

B. Temperate Broadleaf and Mixed Forests (including Qinling Mountains deciduous forests, Daba Mountains evergreen forests, and Sichuan Basin evergreen broadleaf forests). These forests support one of the richest arrays of temperate plant species in the world, with forests that extend eastward from the Hengduan Mountains, across to the northern Sichuan and Sha'anxi Provinces of south-central China. Home to the endangered giant panda (*Ailuropoda melanoleuca*), the Southwest China Temperate Forests also contain a wealth of other species. The Qinling Mountains form an important boundary between two of China's largest watersheds - the Chiangjiang (Yangtze River) and Huang He (Yellow River). This area contains the 7 nature reserves known together as the Sichuan Giant Panda Sanctuaries, and is on the World Heritage List. The fauna of this region includes giant panda, clouded leopard, Chinese muntjac, tufted deer, golden pheasant *Chrysolophus pictus*, Lady Amherst's pheasant *Chrysolophus amherstiae*, Reeves's pheasant *Syrmaticus reevesii*, white eared pheasant *Crossoptilon crossoptilon*, and Temminck's tragopan *Tragopan temminckii*.

The Afrotropic Realm

In Africa, the pheasants are represented by a single species- called the Congo peacock *Afropavo congensis*, which lives in the tropical rainforests of the Congo River basin. Fossil evidence suggests that the two genera - *Pavo* (that contains blue peafowl and green peafowl) and

Afropavo (Congo peafowl) - have diverged from a common ancestral peafowl. The finding of fossil bones of a peacock-like bird (that scientists have termed *Pavo bravardi*) in France and Moldavia (a region now divided between Romania, Republic of Moldova, and Ukraine), shows that the Asiatic genus *Pavo* was more widespread in the past, when dense forests spread over Europe during the Miocene epoch (Mourer-Chauvire, 1989). The distributional anomaly of a single species of an African peacock occurring in the Congo Basin can thus be explained by assuming that a pre-Afropavo form of bird was isolated in Europe during the Miocene, after which it was driven south into western Africa during the onset of colder period (Pliocene epoch) in Europe (Johnsgard, 1999).

The Congo peacock *Afropavo congensis* is found in 2 World Heritage Sites in Democratic Republic of Congo - Salonga National Park (in Congo River basin, Africa's largest tropical rainforest reserve, with *Gilbertiodendron dewevrei* as the dominant tree species), and Okapi Wildlife Reserve. The Congo peacock is also found in the Kokolopori Bonobo Reserve in DRC, which is home to other wildlife such as - bonobos, dwarf crocodiles, forest elephants, Salongo monkey, Thollon's red colobus, bongo, African golden cat, Congo forest buffalo, and birds such as - gray parrot, yellow-legged malimbe, Congo sunbird, and Congo martin (Hance, 2009).

The Ecology of Pheasants: What we know about their habitat and habits

The pheasant species are distributed in basically 3 types of habitats (Johnsgard, 1999). A few of them occur in 2 different habitat types. For example, the blue peafowl *Pavo cristatus* is found in the deciduous forests of the plains, as well as in the lower montane forests in India.

1. Temperate montane forest habitat: (India, Pakistan, Nepal, Bhutan, Japan, China).
2. Tropical and sub tropical forest habitat: (India, Sri Lanka, Southeast Asia, and the Democratic Republic of Congo).
3. Temperate open-country habitat (i.e. grasslands with streams and wooded areas): (China, Mongolia, Korea, Iran, and introduced in former USSR, USA, Europe, Australia, New Zealand, etc.).

The 'temperate montane forest' and 'tropical forest' habitats are further sub-divided into zones of vegetation types depending on the altitude. The different pheasant species have a preference for different vegetation types and altitude ranges. Some of the species migrate seasonally to adjacent vegetation zones, preferring lower elevations in the winters (between December and March). Detailed ecological studies of some of the pheasant species, in their native habitats, are still being conducted, that help us understand the factors affecting their distribution.

Zones in the Montane Habitat

Fig. 1.10 A sketch of the Himalayan monal pheasants (female on the left, and males on the right), by T.W. Wood, in 'Pheasants: Their Natural History and Practical Management' 2nd edition, 1881, by W.B. Tegetmeier.

At daybreak and Siliguri I changed into a tiny toy train with a screaming whistle which vibrated to one's marrow. More miles of dust from which there slowly arose an enormous crimson ball. Then we dived into a forest and began to ascend, winding in and out, around spurs, doubling back, zigzagging, always climbing. We acquired the habit of crawling beneath bridges and a moment later by a convulsive twist of our steel vertebrae (train), passing over them. I fully expected to see the trainlet tie itself into knots before we got through.

First came tall jungle terai with hosts of orchids in flower; then, as we got wonderful vistas of the Teesta and the fast-dropping plains, we reached the zones of tree-ferns and rhododendrons. The heat waves danced over the breathless expanses we had left, while sheer above, waterfalls dashed down through forests - dark, cool and fragrant.

-From *Pheasant Jungles* by William Beebe

The western Himalayan mountain chain extends from Afghanistan, northern Pakistan, north-west India (Jammu and Kashmir, Himachal Pradesh, to Uttarakhand), to western Nepal. A number of rivers originate in the western Himalayas and flow westward, eventually joining the river Indus which flows into the Arabian Sea. Rivers Yamuna and Ganges (and their tributaries) also originate here and flow eastwards across the north Indian plains, and drain into the Bay of Bengal. The forested valleys and floodplains of these rivers are important wildlife habitats.

The various vegetation zones in the western Himalayas have been well studied in the state of Himachal Pradesh, India. The major vegetation zones distributed according to altitude range are (Johnsgard, 1999):

1. 500-1000 m- Subtropical Dry Forest: A mixture of tropical trees and dense under-growth favored by red junglefowl *Gallus gallus.* Indian peafowl *Pavo cristatus* inhabits lower oak forests and agricultural lands.

2. 600-1700 m- Subtropical Pine Forest: Forests dominated by chir pine trees. (In China, this zone has - cypresses, thujas, oak, bamboo, larches, and rhododendrons).

3. 1500-3000 m- Himalayan Moist Mixed Temperate Forest: It is the most important pheasant habitat zone. Temperate tree species, such as oaks, firs, pines, horse chestnut, devdar, spruce, blue pines, blue cherry, walnut, maple, birch, and under-growth of rue, ringal bamboo (*Arundinaria* sp., a slender reed-like bamboo species that grows near streams), magnolias, lilacs, ferns, strawberries, orchids, and mosses are found in this zone.

4. 1500-3000 m- Himalayan Dry Mix Temperate Forest: A zone dominated by Holm oak and edible pine trees.

5. 3000-3400 m- Subalpine Forest: Zone dominated by birches, firs, pines, and under growth of rhododendrons. (In China, this zone has - spruce, holly oak, junipers, ash, liliaceous bulbs, strawberries, and wild onion bulbs).

6. 3400-3500 m- Subalpine Scrub: Dominated by rhododendron trees.

7. 3500-3800 m- Dry Alpine Scrub: Zone of rhododendron and junipers.

8. 4000-4800 m- Alpine Meadows: Grasses and alpine herbs.

Seven species of pheasants are found in Himachal Pradesh. The red junglefowl *Gallus gallus* is mostly found in the subtropical dry evergreen forest below 1200 m, and prefers dense under-growth habitats. Indian peafowl *Pavo cristatus* lives in habitats similar to that of red junglefowl. The kalij pheasant *Lophura leucomelanos* is found in disturbed habitats close to human habitation in lower oak and coniferous forests (Johnsgard, 1999). A study conducted in oak forest patches of Kumaon Himalaya, Uttarakhand, India in 1996-97, has found that the kalij pheasant occupies forested areas with medium tree cover and a tall shrub layer (that occurs mostly along streams) at lower altitudes, while koklass pheasant *Pucrasia macrolopha* occupies forested areas with lower tree cover, and shrub layer of intermediate height at middle altitudes (Shah Hussain et al., 2001). Himalayan monal *Lophophorus impejanus* prefer a less dense ground cover. For cheer pheasants *Catreus wallichii*, the presence of dense grass appears to be an important habitat attribute. The western tragopan *Tragopan melanocephalus* is associated with middle altitude meadows and temperate forests.

Tree cover and slope are important to pheasants because they determine the vegetation characteristics, or the type of food available at ground level. Greater slope also prevents access by people and livestock, and perhaps also facilitates escape from predators by downhill running. Distribution of pheasants in China indicates a similar vertical segregation of pheasant species in vegetation zones (Johnsgard, 1999).

Fig. 1.11 Alpine forest and grassland habitat of the Himalayan monal in the state of Uttarakhand in northern India. Photo by Bijoy Venugopal, 2009.

Fig. 1.12 Blood pheasant males and females in Khumbu, Nepal, by Peteris Erglis, 2006.

Fig. 1.13 A photo of koklass and kalij pheasant habitat in Water Catchment Area Sanctuary near Shimla, Himachal Pradessh, India, containing wet mixed temperate forest interspersed with streams and grasslands, by Anita Chauhan, 2010.

Zones in the Tropical Habitats

Fig. 1.14 A sketch of great argus pheasants (female on the left, and males on the right), by T.W. Wood, in 'Pheasants: Their Natural History and Practical Management' 2nd edition, 1881, by W.B. Tegetmeier.

Thus I made my ascent to the summit of the great mountain range (in a motor lorry), amid a continuous whirl of choking dust which quite obliterated the scenery. I might have imagined myself caught up in a cloud, as worthy biblical characters were wont to be, only I am sure they were spared the odor of burning oil and rubber, and their ears were not assailed with a syncopated obstruction in the brake mechanism which, before the end of the trip, vied with the efforts and effect of any brain-fever bird.

When we emerged from our cloud and excavated our eyes we found a wonderland, a little rose-covered dak cottage with an immaculately saronged Cinghalese in attendance, and tiffin of curry and tea. This was Semangko Pass – the Darjeeling and the Simla of the Malay Peninsula.

Semangko Pass struck into my memory as the most beautiful of the tropical mountains of the East where I strove to match my senses against those of the jungle pheasants. The dak was perched on a little flat saddle at the very crest of the ridge, scarcely less than three thousand feet above the blue waters of the Indian Ocean. On all sides the sharp-toothed mountains rose still higher, steep but jungle-clad, cutting the sky into all sorts of irregular bits of glory.

The days were wonderful, and the alternations of sun and wind were as exciting as the discovery of the strange Malayan beasts and birds. The sun rose softly – no breeze moved cloud or leaf, and even the light came at first moderately, indirectly, reflected from the higher peaks, or heliographed from the mirror of a half-hidden, distant waterfall. In early afternoon – one never knew just when – the faintest of breezes sifted down and blurred the lacery of tree-fern shadows. The wind was cool and soon strengthened, and by night it was surging through the gap, flowing from the cold summits down to the hot, humid valleys.

Day after day one reawakened to the sense of tropical surroundings from a conviction of a northern autumn, with the wind full of swirling leaves and the fronds soughing with the same sad cadence as the needles of scented pines of the northland.

The first night I listened to the strange sound of wind in the eaves of the bungalow. And then I slept, and was awakened by the distant, faint chorus of wa-was, the long-handed gibbons, a sound as thrilling, as full of old memory, as the morning chant of the red howlers in the South American jungles.

The liana-draped trunks and the majestic jungle trees were the finest in all the East, second only to those of Amazonia, but the tree-ferns were beyond words – tall, graceful, with great unfolding fronds half-clenched, swathed in wool of richest foxy-red. Here, in this maze of mountain jungle, through its autumnal days and its wild, tropic nights, lived two splendid races (species) of birds. One was the bronze-tailed peacock pheasant, the other the giant ocellated argus. Each was a challenge to my utmost effort. Neither had been seen by a white man; of neither had we any facts of home or courtship, food or foe.

-From *Pheasant Jungles* by William Beebe

Altitudinal zonation of vegetation, similar to that in the western Himalayas, also occurs in the tropical evergreen and semi evergreen forests in Southeast Asia (Johnsgard, 1999). We can take Malaysia as an example country to illustrate this. Malaysia consists of 13 states, 11 in the Peninsular Malaysia (attached to the Asian mainland), and Sarawak and Sabah on the island of Borneo (East Malaysia). 'Lowland tropical rain forest is the principal original forest type in Peninsular Malaysia. The rain forest is rich in trees of the family Dipterocarpaceae, and can be subdivided into lowland forest (below 300 m), and hill forest (300-1000m). Above 1000 m, tropical montane evergreen forest is found. Sabah and Sarawak were originally covered in forest, including tropical lowland rain forest, and lower and upper montane forests. Forest cover in Sabah has shrunk from 86% of the land area (in 1953) to 45% of the land area (in 1985) due to commercial logging' (Watkins et al., 1997).

Both Peninsular Malaysia and northern Borneo are recognized biodiversity hotspots. By 1993, 293 mammals, 1200 birds, 171 amphibians, 294 reptiles, 449 freshwater fishes, 54 swallowtails, and 15000 higher plants were documented. Malaysia is home to many species of conservation and economic interests, including tiger, elephant, orangutan, siamang gibbon, and proboscis monkey, and important birds, such as crested argus pheasant *Rheinardia ocellata*, Bulwer's pheasant *Lophura bulweri*, and helmeted hornbill; and has economic species of medicinal plants, and timber trees of family Dipterocarpaceae for which Borneo is the world distribution center. Sarawak and Sabah have many of the endemic species of the island of Borneo (Watkins et al., 1997).

The Malayan evergreen forest is characterized by an abundance of Dipterocarp *Shorea* species, high tree species diversity, and a tall forest profile (with a multi layered vegetation canopy). 8 species of pheasants occur in Malaysia, out of which the red junglefowl *Gallus gallus* and green peafowl *Pavo muticus* have a wide distribution in Southeast Asia (though the green peafowl is extirpated from Peninsular Malaysia and 'Endangered' in the rest of its range). 3 types of forests are found in this area (altitude range 0 to about 4095 m) (Johnsgard, 1999):

1.   The lowland forest (below 300 m) consists of lowland- and hill Dipterocarp subtype of forest. 3 of the species of forest-dwelling pheasants are found in the lower-Dipterocarp community: crested fireback *Lophura ignita*, crestless fireback *Lophura erythrophthalma*, and Malayan peacock pheasant *Polyplectron malacense*. The crested fireback is found only on level ground near rivers, and the latter 2 are found in sloping or steeply sloping habitats in the forest. The great argus pheasant *Argusianus argus* is found in most lowland- and hill-Dipterocarp forest sites, but is absent from heavily disturbed and fragmented logging sites.

2.   The lower montane forest (300-1000 m) consists of upper-Dipterocarp and Oak-Laurel subtype of forests. The crested argus *Rheinardia ocellata* is found in the hill-Dipterocarp and lower montane transitional forests.

3.   The upper montane forests (above 1000 m) consist of a montane Ericaceous community (heath family containing rhododendrons, etc). The Rothschild's peacock pheasant (a.k.a. mountain peacock pheasant) *Polyplectron inopinatum* has been observed only in locations above 900 m (and up to 1800 m), typically on steep slopes with vegetation and exposed rocks.

Temperate Open-Country Habitat

This type of pheasant habitat consists of the following environments – broad wooded river valleys close to grasslands, foothills, overgrown edges of rivers, hilly areas close to large cultivated fields having small bamboo groves or low pine thickets, and flat lands cultivated with rice, wheat, etc (Johnsgard, 1999). The ring-necked pheasant *Phasianus colchicus* and junglefowl *Gallus* sp. are found in such habitats.

General Habits of Pheasants

It was a memorable moment when I and my string of heavily laden Tibetan women and their no heavier laden husbands, turned our backs on Darjeeling and headed northward into the mysterious Himalayas. I watched the two and thirty bent figures, and as the men and buffaloes in the plains had seemed like mere coagulated dust wraiths, so these Tibetans with their coarse, harsh garments and broad, squat figures might well be bits of their native crags. They seemed "rocks that walked", as if great boulders of hardy granite, weathered and worn, carried down from the heart of the glaciers of Everest and Kinchinjunga, had rolled tumbling down the icy slopes, and before they came to a stop, had been endowed with human form and life by some whimsical god of the mountains.

I decided to make my first stop at the dak bunglow of Jorepokri, at about seven thousand feet elevation, almost the same as that of Darjeeling. My caravan reached it by winding trails shadowed by oaks and maples. Here, for days, I explored the dense forest alone, and studied the habits of the black-backed kalij pheasant, a beautiful steel-blue, white-breasted bird with the hen clad in red-brown and russets.

-From *Pheasant Jungles* by William Beebe

The daily routine of pheasants generally follows a simple pattern. After emerging from the roost in the morning, the birds drink water from a stream or a rain-puddle, and feed for about 2-3 hours, before retiring to shelter during the middle of the day, reappearing to drink and feed in the late afternoon. Feeding might continue till dusk. Very little is known about the daily activities of forest pheasants. Radio-tracking of a few Malayan peacock pheasants *Polyplectron malacense* revealed that the birds were mobile for much of the day - there was no midday resting period (McGowan, 1994).

Fig. 1.15 A male kalij pheasant perched on a deodar tree in the Chhota Shimla locality of Shimla hills, Himachal Pradesh, India. Photo by Anita Chauhan, 2010.

Male great argus *Argusianus argus*, however, was found spending some time sitting on low branches near dancing grounds during the day. For most months of the year, many species, and primarily the males, start the day by calling. In open country, this calling is often from atop rocky outcrops or perches in trees or bushes. Some forest species, such as the Malayan peacock pheasant and the Satyr tragopan *Tragopan satyra*, are known to call from elevated branches 10-20 m above the ground (McGowan, 1994).

Fig. 1.16 A male blue peafowl perched on a mango tree in the afternoon in New Delhi, India. Photo by Anita Chauhan, 2011.

Fig. 1.17 Blue peafowl males in the spotted deer enclosure in the National Zoological Park, New Delhi, India, by Anita Chauhan, 2009.

In closed forest habitats, where females can be hidden by vegetation, males do not need to guard them, and as a result many more species are solitary, for instance, the Palawan peacock pheasant *Polyplectron napoleonis*. There are exceptions to this, such as the crested fireback *Lophura ignita*, an inhabitant of primary rainforest, which can be seen in flocks of up to 10 individuals. 'Several pheasant species in Southeast Asia are not known to associate (as pairs or groups) for any length of time, the two sexes meeting only to copulate, as in the case of great argus *Argusianus argus* and crested argus *Rheinardia ocellata*. In contrast, many open country species are often found in larger groups, at least outside the breeding season, and often in pairs while breeding. In the monogamous cheer pheasant *Catreus wallichi*, it is reported that several pairs will join together to form a single flock during the winter. There are also reports of eared pheasants *Crossoptilon* sp. in China occurring in flocks of up to 100 birds during the winter' (McGowan, 1994). So, the degree to which some species are gregarious varies through the year. While during the non-breeding season a number of species are found in groups, these tend to disperse in the breeding season. A study in North America has found that during winter, ring-necked pheasants *Phasianus colchicus* form temporary flocks, and individuals move about and feed together as a relatively coherent unit with shifting membership. Males and females sometimes feed together, but also often form unisexual groupings. Roosting groups during winter vary from 2 to 24 birds. With the onset of spring, male groups occurring in pairs, trios or more, start dispersing to establish their territories. The mixed sex groups disperse gradually to harems of hens, with each harem dominated by a single male (Johnsgard, 1999).

In India, researchers from the Aligarh Muslim University (Uttar Pradesh, India) have conducted a preliminary study on 4 species of pheasants (white-crested kalij *Lophura leucomelanos*, koklass pheasant *Pucrasia macrolopha*, Satyr tragopan *Tragopan satyra*, and Himalayan monal *Lophophorus impejanus*) found in the Kumaon region in the state of Uttarakhand in western Himalayas. The researchers have found out interesting details about the habitat preferences of the 4 species (see a further discussion under 'Pheasant Calls and Vocalizations' in this chapter), and have collected some data about their social group size and composition (Shah Hussain and Sultana, 2013). Further studies on the sociality of these pheasants are needed to fully understand it.

Activities of Pheasants

Like most other bird species, pheasants also perform activities such as preening and dust-bathing to keep their feathers in good condition. During preening, the bird takes the oily secretion of the uropygial gland at the base of its tail, and spreads it over its feathers with its beak. Dust-bathing involves the bird lying on its side in loose dry soil, and scooping in the dust with its wings. This 'powdering' action helps in getting rid of parasites. 'These activities and behaviors are present in pheasant species from the time of hatching or shortly thereafter, and include locomotory behavior (walking, running, jumping, hopping, and flying), comfort behavior (scratching, preening, and dust-bathing), and feeding behavior. The young chicks are precocial (able to feed themselves at hatching), and often are able to fly short distances within a week or so after hatching' (Johnsgard, 1999).

Fig 1.18 A. Cheer pheasant preening in Himalayan Bird Park, Shimla, H.P., India. B. Peahen preening, New Delhi, India. Photos by Anita Chauhan, 2010.

The sleeping posture, of red junglefowl and other galliform birds, involves tucking the beaks into the scapular (shoulder) feathers. This posture is not fully attained until about the 14th day after hatching, when feathers in this region have grown sufficiently to hold the beak in place. Early escape behavior includes the alert posture, head shaking, running, and squatting. Hopping, leaping, pecking, and kicking towards the opponent during leaping, develop from 1-3 weeks of age. Soon, social peck-orders become established, and young males (called cockerels) generally become higher in rank than young females (called pullets). A more or less linear social hierarchy develops, with the linearity probably resulting from the deferential rates of development in the individual birds. As male chicks become older, they exhibit social behavior patterns that become eventually integrated into courtship and reproductive behavior (Johnsgard, 1999).

Feeding: What do the pheasants eat?

Fig. 1.19 Two male Himalayan monal pheasants in Uttarakhand, India. Photo by Bijoy Venugopal, 2009.

We think of a humming bird as quite the most brilliant and colourful creature in the world – a strange little being with the activity and bulk of an insect, the brain of a bird, and the beauty of an opal. Imagine one of these, shorn of its great activity but enlarged many times, and one has an impeyan pheasant (monal) of the Himalayas. Beneath, it is black as jet; its crest is a score of feather jewels trembling at the extremity of slender bare stalks. But its cloak of shimmering metal is beyond description, for with each change of light the colors shift and change.

When the shadow of a cloud slips along the mountain slope the impeyan glows dully – its gold is tempered, its copper cooled, its emerald hues veneered to a pastel of iridescence. But when the clear sun again shines, the white light is shattered on the impeyan's plumage into a prismatic burst of color.

My eye caught a trembling among the maiden-hair fern, and I swung my glass (binoculars) and brought a full-plumaged impeyan into the field (of view). The dew and the soft light of early dawn deadened his wonderful coat. His clear brown eyes flashed here and there as he plucked the heads of tiny flowers from among the grass.

For fifteen minutes nothing more happened; then for the space of an hour, impeyans began to appear singly or in pairs, and once three together. Finally, fourteen birds, all cocks in full plumage, were assembled. They gathered in a large glade which already showed signs of former work, and there dug industriously, searching for grubs and succulent tubers. They never scratched like common fowl, but always picked with their strong beaks. Every 3 or 4 seconds, they stood erect, glanced quickly about, and then carefully scanned the whole sky. It was easy to divine the source of their fear – the great black eagles which float miles high like motes. The glittering assemblage fed silently, now and then uttering a subdued guttural chuckle.

When the sun's rays reached the glade, the scene was unforgettable: fourteen moving, shifting mirrors of blue, emerald, violet, purple, and now and then a flash of white, set in the background of green turf and black, newly upturned loam.

-From *Pheasant Jungles* by William Beebe

Pheasants vary individually, geographically, and taxonomically with respect to food and foraging adaptations. All pheasants are predominantly vegetarian, and most are adapted to seed-eating. Some species, such as koklass pheasant *Pucrasia macrolopha*, consume a high proportion of green foliage, whereas others like the monals *Lophophorus* sp., dig for subterranean vegetable matter i.e. roots and bulbs (Johnsgard, 1999).

All pheasants primarily forage on the ground. Their sharp beaks, stout toes, and strong claws are well adapted for digging and scratching. There is some variation in beak shape among the pheasants. The monals, for example, have unusually well developed beaks for digging out foods from below the ground surface. Roots, bulbs, tubers, sub-soil insects, and other invertebrates are excavated and consumed. A few pheasant species, such as the blood pheasant *Ithaginis cruentus*, have comparatively weak beaks, and primarily eat greens and fruits. All species have well-developed crops (storage organ in the alimentary canal) for temporary storage of food, as well as muscular gizzards for grinding of hard food materials with the help of grit (Johnsgard, 1999).

In all the birds of family Phasianidae, a high proportion of live animal food, especially insects, is consumed during the first few weeks of life, and this ratio of animal to plant foods decreases during ontogeny. From the information available on many of temperate-zone species, it is concluded that by the time the birds are adults, about 90% of food intake comes from plant sources. Birds whose primary food is mostly leaves (for example, tragopans and eared pheasants), may eat about 15% dietary protein, whereas highly insectivorous tropical species (for example, argus and peacock pheasants) may eat up to 40% protein (Johnsgard, 1999).

Dr. Beebe states in his monograph on pheasants that tragopans are omnivorous foragers but tend to specialize in leaves and buds. Beebe found that western tragopans *Tragopan melanocephalus* forage on newly sprouted leaves. They also eat roots, flowers, insects, acorns, seeds, and berries of various kinds, but in small amounts as compared with leaves (Johnsgard, 1999).

Beebe also found that Satyr tragopans *Tragopan satyra* eat leaves, flowers of the paper laurel and rhododendrons, and insects such as small earwigs, ants, spiders, cockroaches, and centipedes. Satyr tragopans also eat bamboo shoots, ferns, onion-like bulbs, wild fruits, and rhododendron seeds. Tragopans tend to favor scarce but potentially high quality foods such as insects or fallen fruits. In captivity, the birds are largely vegetarians, with an emphasis on fruits and berries (Johnsgard, 1999).

Fig. 1.20 Photo of a male red junglefowl scratching amidst rhino dung-pile for beetles in Kaziranga National Park, Assam, India, by Aditya Singh, March 2009.

Beebe reported that, like most pheasants, Satyr tragopans usually confine their foraging activities to early morning and late afternoon hours. On dull and cloudy days, however, they may forage at irregular intervals. Satyr tragopans typically forage on the open edges of the forest, and scratch the leaf litter or under-growth; or they may feed in low trees and bushes to obtain petals, buds, and berries. When feeding in jungle under-growth, Satyr tragopans apparently concentrate in a few likely spots, rather than scratching superficially and randomly over a wide area as in the case of kalij pheasants *Lophura leucomelanos* (Johnsgard, 1999).

Beebe has also summarized the foods of the Himalayan monal *Lophophorus impejanus.* He suggests that terrestrial insects and tubers form their chief diet, but the specific food varies by locality. Wherever snow does not cover the ground, the birds spend a lot of time digging with their beaks for tubers, roots, and subterranean insects. In autumn they are said to forage mainly on insect larvae found under decaying leaves. They also feed on leaves and young shoots of various shrubs and grasses, as well as acorns, seeds, and berries. Although the Himalayan monals may be seen in wheat and barley fields in winter, they seek roots and insects rather than the grain. Edible mushrooms, wild strawberries, currants, and the roots of ferns are also included in their diet. The foraging behavior of the monal is very distinctive. The birds do very little digging with their feet, but instead pick at the earth with their shovel-like beaks, sometimes digging holes as deep as a foot. Himalayan monals typically forage in small groups (Johnsgard, 1999).

Tropical forest species generally appear to have less stout beaks and feed by turning over leaf litter, either with the beak alone, or using their feet to scratch away as well. Birds then peck at the exposed soil surface, picking up various foods, including invertebrates such as giant forest ants (ant genus *Camponotus*). In Malaysia, the crested fireback *Lophura ignita* and crestless fireback *Lophura erythrophthalma* - species that often form groups - tend to scratch at the leaf litter to eat the invertebrates that live under this layer. These invertebrates are often small colonial species, and when one of the birds finds food (also including clusters of fallen fruit), the other members of the flock come to feed beside it. Solitary species, such as the great argus *Argusianus argus*, do not tend to scratch the ground, and simply pick up bugs from the surface of the litter. The peacock pheasants *Polyplectron* sp. appear to favor a diet of insects (McGowan, 1994).

Foraging is, therefore, an important part of the daily routine of the phasianids. Among pheasant species that forage in the open habitats, it most typically occurs in the early morning or late afternoon. This is often because of the midday heat of the tropics and the subtropics, but as in case of the Chinese monal *Lophophorus lhuysii*, the presence of potential predators is a major determinant of feeding times (McGowan, 1994). As mentioned earlier, pheasants perform the ecological function of seed dispersal, and control of insect and small reptile populations in the ecosystem. Pheasants feed on a variety of plants; their occurrence in a habitat indicates a diverse flora.

Fig. 1.21 A painting of a pair of Blyth's tragopan *Tragopan blythi* (male in the foreground) by Archibald Thorburn in 'A Monograph of the Pheasants' by William Beebe (created between 1918-1922).

Fig. 1.22 A. Rhododendron tree in bloom in Chhota Shimla, Shimla hills B. Berries of *Principia*, Shimla hills, Himachal Pradesh, India. Photos by Anita Chauhan.

Fig. 1.23 A male Himalayan monal feeding under rhododendron bushes, Uttarakhand, India. Photo by Bijoy Venugopal, 2009.

Pheasant Calls and Vocalizations

To wake up in a tent and look out is good; to sit up in one's blanket cocoon in a hammock and see the jungle dawn is better; but best of all is opening one's eyes in a houseboat bunk, and without further movement seeing water and jungle and sky, and the exciting early morning doings of fish, crocodiles, birds, and monkeys. One feels as yet unburdened with a human frame; and for an hour I am only a pair of disembodied eyes, which search and record, begrudging even the interruption of blinks, and viewing all through fresh-colored, wide-angle vision.

As one wakes up slowly from slumber, so came the dawn, gradually, in these tropical lowlands. The glare of the sudden leap of the sun above the horizon was dimmed, delayed, diluted, by the thick morning mist – mist whose grayness I loved to think of as the exact shade of "elephant's breath". As I looked out over the side of the boat, the swift current became more and more distinct through the fog, which drifted slowly downward like a sluggish, aerial river flowing gently over the denser one below. When the light grew and the mist lifted and frayed upward, a brown line quartered the fore-glow in the sky and masses of foliage took shape and color beyond the sand-banks. Here and there white-barked trunks gleamed like ghosts, the saturated air was heavy with the odor of plume-blossoms, and the eddies were filled with their petals. A pair of great hornbills crossed high overhead, hidden by cloud-mist, but registering every wing-beat in a loud, deep, *whoof! whoof!* Bulbuls burst into song, drongos sent down their hoarse cries from the tree-tops, with showers of drops which pattered on my cabin roof.

Another veil of mist was drawn aside, and I sat up, breathless and tense, for on a sand-bar upriver and up-wind four great black forms became dimly visible – giant, statuesque sladang, the biggest bull standing at least six feet at the shoulder. Even against the pale sand, their cream-colored stockings showed clearly, and their magnificent curved horns lay far back as they stood with nostrils outstretched toward me, striving to make out by sight what the wind refused to explain. We seemed harmless – some huge tree stranded during the night; but with wilderness folk, vague suspicion is interpreted as proven danger, and the wonderful jungle cattle, still headed our way, moved slowly through the shallows around and behind an arm of foliage.

The other end of the sand-bar held for me even greater interest. Resting my stereo glasses on the edge of the bunk, I was fairly in the midst of five green peafowl. They had me under surveillance, but were too confident of their powers to think of leaving. Two had sweeping trains which cleared the damp sand as they walked. Now and then a bird stood quite erect and flapped his wings vigorously, to rid the feathers of excess of moisture. I could even see the others shake their heads as the drops flew over them. Two young of the year were very active, running about, chasing one another, or stopping to scratch among the gravel.

A passing log drew the attention of the peafowl, and they all stood motionless, watching it, until they were certain it was wood, not crocodile. The sun shone brightly for a moment, and the mists swirled away, showing distant hills. Peal after peal of rollicking laughter came from a family of serious-faced wa-was. Then a rush of wind and fog blotted out the sun, and a sudden shower pitted the smooth water. From the depths of this renewed twilight rang the piercing unrestrained call of the wild peacock.

-From *Pheasant Jungles* by William Beebe

Birding enthusiasts love to hear the delightful sounds of bird-calls, as much as they like to behold bird plumage and antics. But where can one hear pheasant calls? The more adventurous among us like to visit nearby sanctuaries and national parks to sight the rare forest species. Most Protected Areas have walking trails that can be used by tourists to explore the wildlife habitat. An early morning walk in a pheasant habitat (such as the Chail Wildlife Sanctuary, Water Catchment Area Sanctuary, and Daranghati Sanctuary in Himachal Pradesh, India) is the best way to sight/hear cheer pheasants *Catreus wallichi*, koklass *Pucrasia macrolopha*, white-crested kalij *Lophura leucomelanos*, and western tragopans *Tragopan melanocephalus*. These walking trails are also used by wildlife scientists to survey pheasant populations. A population survey involves walking along the trails to flush out the pheasants (or to record their calls, or the presence of droppings or feathers), which are enumerated per hour of walking. We will read more about this in Chapter 4.

Fig. 1.24 A walking trail in Water Catchment Area Sanctuary, Shimla, Himachal Pradesh, India. Photo by Anita Chauhan, 2010.

Some species, like the Himalayan monal *Lophophorus impejanus*, are found close to forest-edge village habitats during the dead of winters, when snow drives them down to fields and orchards in search of food. Kalij pheasants *Lophura leucomelanos*, on the other hand, live close to villages or suburbs even during their breeding season. In a protected deodar forest containing a stream in the Chhota Shimla locality of Shimla Hills, Himachal Pradesh, kalij pheasants can be seen and heard at dawn and dusk just meters from human habitation. The rich vegetation near the stream (including *Rosa, Impatiens, Principia, etc.*) and the springtime growth of edge-habitat plant species like 'Kungshi' (*Girardinia heterophylla*), provides cover to the Kalijs during nesting and feeding. Such habitats have, however, shrunk in Shimla hills due to construction and road building activities; and areas in the city that are notified as 'Green Belt' by the Shimla Municipal Corporation offer the main refuge to wildlife (Chauhan, 2012).

That kalij pheasants require a dense and tall ground cover in forests has been demonstrated in a study of the ecology of kalij and koklass *Pucrasia macrolopha* in 23 oak forest patches of Kumaon Himalaya, Uttarakhand, India in 1996-97. It was found that the kalij occupies forested areas with medium tree cover and a tall shrub layer (that occurs mostly along streams) at lower altitudes, while koklass occupies forested areas with lower tree cover and a shrub layer of intermediate height at middle altitudes (Shah Hussain et al., 2001). In the urban areas in other Himalayan cities too, it is important to conserve such high diversity habitats along the streams, and to notify buffer zones of 100 feet or more on the sides of the streams to prevent any construction activity. Similarly, buffers should be notified by the respective municipalities for wetland, or river bank habitats in the plains, in order to: (a) conserve them as 'Green Infrastructure' that provides ecosystem services, nature education, and recreation opportunities, and (b) conserve representative flora and fauna of the city's natural ecosystem that imparts a sense of identity to urban dwellers. Participatory programs (such as 'nature walks', natural history societies, and newsletters) to involve people in conservation and monitoring of natural habitats should be provided (Chauhan, 2012). This is further elaborated in Chapter 6.

Bird calls serve as a territorial claim, and to attract females. "Crowing" is the best known of male calls, although occasionally, egg laying hens may also crow. Crowing by pheasants is most often performed at dawn and dusk, and especially during spring, but may occasionally be heard throughout the year. The crowing rate of a male seems to be positively related to his position in the peck-order. Fog or mild rain does not seem to have an effect on crowing rates (Johnsgard, 1999).

Fig. 1.25 Kalij pheasant pair chorus together in the breeding season at morning or evening hours. One pair was seen calling at 5 PM while perched 6-7 m high on a deodar tree in Chhota Shimla, Shimla hills, Himachal Pradesh, India. Photo by Anita Chauhan, August 2010.

During the spring (breeding season), males of many species make the vocalizations that are heard only at this time of the year and include advertising and courtship calls. It is the advertising calls that have tended to attract most attention, as they are typically loud and far-carrying. In open country habitats, many males move to conspicuous positions to make these calls. The daily pattern of calling during this period typically has a peak around dawn and, in some species a secondary peak at dusk. For example, in the solitary, forest-dwelling Malayan peacock pheasant *Polyplectron malacense* and great argus *Argusianus argus*, there appear to be 2 types of advertising calls. In the former species, the calls, "tchorrs" and whistles, convey for up to 300 m through the forest undergrowth. The calls verbalized as "tchorrs" are harsh-sounding, and each call contains a single note of narrow frequency range (below 2 KHz), and lasts between half a second and a full second. Whistles are melancholy double note calls, the 2nd note of higher frequency than the 1st. They are also of narrow frequency range (below 1.2 KHz), and are lower-pitched than the "tchorrs". Both calls are often repeated many times, and both are sometimes heard in the same bout, in which case "tchorrs" are always emitted first. It is likely that the number and type of calls given by a male indicate his reproductive condition. The two advertising calls of the great argus are loud, and are often given from near the top of low hills. These calls are amongst the most evocative sounds of the Southeast Asian rain forest, and they can be heard as much as a kilometer away (McGowan, 1994).

In the Himalayas, male koklass pheasants *Pucrasia macrolopha* chorus in the early mornings from November to June, just as the first light appears in the sky. Chorusing appears unaffected by the weather, and usually lasts about half an hour, finishing before sunrise. The cheer pheasant *Catreus wallichii*, which also occurs in the Himalayas at mid-altitudes, choruses at dawn as well as dusk. Both sexes of this monogamous species participate in the chorus, the paired male and female usually calling together in such a way that it sounds as if only one individual is giving a very complex call. First calls are usually given an hour or less before sunrise, and may, on occasion, last up to an hour after it. In the evening, the chorus typically lasts from up to an hour before, to half an hour after sunset (McGowan, 1994).

The conspicuous calls of a pheasant are often the first indication that a particular species is present in an area, i.e. it is a harmless method of determining the presence and identifying a species. Hence, calls are used as indicators during population surveys of species which are little known or difficult to sight – including, western tragopan *Tragopan melanocephalus*, koklass pheasant *Pucrasia macrolopha*, and cheer pheasants *Catreus wallichi* (McGowan, 1994).

Scientists have categorised pheasant calls into various kinds, based on their duration, pulse rate, call loudness, and the duration of intervals between calls. For example, scientists have described 15 different types of calls of ring-necked pheasants *Phasianus colchicus*, including 3 characteristic of chicks, 6 limited to females, 3 limited to adult males, and 2 characteristic of adults of both sexes (Johnsgard, 1999). The types of calls studied in domestic fowl and pheasants include:

1. Ground Predator alarm call
2. Aerial predator alarm call
3. Brood caution call

4. Threat calls
5. Alert call
6. Flight call
7. Brood-gathering call
8. Hissing calls
9. Pecked call
10. Pre-copulatory call
11. Nest-defense calls
12. Crowing/territorial call

Pheasant sightings and calls are a tourism attraction, and bird-watchers from all parts of the world undertake journeys to the forests of Asia to see the rich avifauna of the region. Birding tours are popular in China and Southeast Asia, and are also starting to emerge in India.

Home Ranges and Migrations

Some of the pheasant species are known to migrate locally, spending summer months high in the Himalayas (or in the case of Chinese montane pheasants, high in the mountain ranges of China), and descending to spend winters at lower altitudes.

The western tragopan *Tragopan melanocephalus* in Himachal Pradesh breeds at elevations from 2400 – 3600 m, and winters at about 1350 m. In winter, western tragopans are found in the thickest parts of the oak, chestnut, morenda pine forests having a dense under growth of ringal bamboo. During the breeding season (March – June), they are found in the higher parts of the forest, up to the zone of birches and white rhododendrons and almost up to the upper limits of the forest. During autumn, the families gradually begin their descent to wintering areas. In comparison, the Himalayan monal *Lophophorus impejanus* in H.P. concentrates mainly between 2000-3000 m during January to March, and mostly above 3000 m during September to October (Johnsgard, 1999).

Using Radio-Telemetry to Determine Home-ranges

The home range of an animal is the spatial extent of its movement during the course of its everyday activities. It is determined by placing a radio-transmitter on an individual, and collecting data of the location points over a period of time.

Radio telemetry enables the biologist to follow the daily movements and activities of a bird without disturbing it. The technique has been used with great success on pheasants (Dowell et al., 1992). For instance, 'in a radio-telemetry study of the Cabot's tragopan *Tragopan caboti* in China, it was found that the extent of home ranges increases in spring season. Dominant males have stable home ranges, whereas females wander over large areas, sometimes associating with a male within his range. Female home ranges are largest during the pre-breeding season, decrease during nesting, and increase again after hatching' (Johnsgard, 1994). Similarly, studies on the red junglefowl *Gallus gallus* in north-central India indicate a daily movement of flocks in an area averaging about 200 m in diameter.

A number of studies have been conducted on the ring-neck pheasant *Phasianus colchicus* to find out its home range, and have indicated a small seasonal and daily mobility pattern. As mentioned earlier, the ring-necked pheasants tend to form large groups (2-24 birds) in winters. During spring, these groups give way to harems dominated by a single male. A study conducted in Sweden revealed that the male ring-necked pheasants establish their breeding territories during a two week spring period, and older males return to their previous territories in the following year. During spring, both the sexes move from their wintering habitats in the woodlands to more open habitats, for nesting. Females move about in various male territories, eventually establishing nesting range within one male's domain. Older females typically return to their previous year's mates and their respective territories (Johnsgard, 1999).

Young females disperse farther than adult. Younger males also move farther than older, previously territorial males. About 20% of the yearling males are not able to establish breeding territories and constitutes a 'floating' male population, until subsequent spring seasons (Johnsgard, 1999).

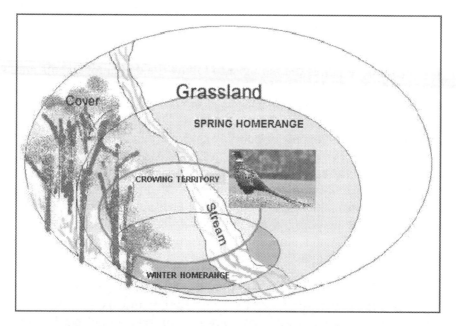

Fig. 1.26 A diagram showing the winter and spring home ranges of a common pheasant in its open country habitat.

Fig. 1.27 A Svalbard rock ptarmigan affixed with a 20 g satellite tag on its back. The Svalbard rock ptarmigan (*Lagopus muta hyperborea*) is a galliform bird of the family Tetranoidae found in Norway. Photo by Eva Fuglei, 2009.

During winter, the daily movements between food sources and cover are usually less than 0.40 kilometers. As home ranges increase in size in spring, there may be an overlap of home ranges of male pheasants. In one study, spring activity in male ring-necked pheasants was found to concentrate on localized areas of less than 1.30 km² within the home range. Such areas have been called crowing territories (or breeding/defended territories). In the male's crowing area, the male does his most intense crowing, the harem gather within, and courtship and displays are most prevalent. Another study has found that these well-defined territories range in size from 0.5 to 4.0 hectares (Johnsgard, 1999). [The center of a standard sports stadium with an athletics track is about 1 hectare in area].

Predators

In their natural environment, the populations of pheasants are controlled by a variety of mammalian, avian, and reptilian predators. In western Himalayas, northeast Himalayas, and Southeast Asia, the following predators are found - fox, marten, dhole, jackal, leopard, eagle, Himalayan black bear, weasel, and snow leopard (excluding in SE Asia); tiger, clouded leopard, sun bear, and python (in NE India and SE Asia only); jungle cat, civet cat, blue magpie, jungle crow, and mongoose. In China - falcon, goshawk, and eagle owl are found. In Taiwan - ferret-badger is the predator. In Japan - hawk, crow, raven, magpie, snake, feral cats, and feral dogs prey on pheasants and their nests.

In Himachal Pradesh, India, the western tragopan *Tragopan melanocephalus* occurs in company with 4 other species of pheasants. A substantial number of predatory mammals inhabit Himachal Pradesh [3 canids (jackal, dhole, and wolf), 2 bears (brown bear and black bear), a weasel (*Mustela sibirica*), a marten (*Martes flavicular*), a civet (*Paguma larvata*), a cat (*Felis chaus*), red fox, and 2 leopards (*Panthera pardus* and *Uncia uncia*)], most of which might represent varying levels of threat to the pheasants. The Himalayan monal *Lophophorus impejanus* also seems to be a prey for eagles - it has been observed to frequently gaze up into the sky while foraging in exposed areas. Various predatory mammals undoubtedly affect the monal as well, but specific information is lacking. The raven is a nest predator for this species (derived from Johnsgard, 1999).

Fig. 1.28 Photo of a red fox in the high altitude cold desert of Hemis National Park, in Trans Himalayan Jammu and Kashmir, India, by Aditya Singh, March 2010.

Hundreds of field-studies have been done on ring-necked pheasants *Phasianus colchicus* in grassland/agricultural field habitats in Michigan and Wisconsin, USA (where it is an introduced species), using radio transmitters. On an average, the studies attributed 40-50% cases of pheasant predation to avian predators and 35% to mammalian predators. Foxes (red and grey) are believed to be responsible for the majority of mammalian predation cases, whereas smaller numbers are attributed to predators such as cats, dogs, minks, and weasels. Among avian predators, the great horned owl, the red-tailed hawk, the Cooper's hawk, and the northern harrier are significant sources of predation (Johnsgard, 1999).

Reproductive Biology: Courtship and Nesting in Pheasants

Many species of animals have evolved complex reproductive strategies to ensure reproductive success in diverse ecological conditions. For example, lions form a pride consisting of 1 or 2 males and several females; and deer form a harem, consisting of a single dominant male and several females. The diverse reproductive patterns seen in nature include differences in the male's participation in post-fertilization aspects of reproduction, with consequential variations in the duration of pairing.

In most species of pheasants, the breeding season is distinctly seasonal. This means that breeding is typically associated with the spring months in the temperate regions or with the wet season in more tropical areas. The seasonal timing of egg-laying is perhaps set by some environmental controls such as photoperiod, the duration of sunlight hours per day. As mentioned earlier, the social groups formed during the non-breeding season gradually give way to the social organization typical of the species in the breeding season.

Mating Systems

A number of interesting observations/suggestions have been made by scientists, on the mating systems of pheasants relative to ecological aspects of their habitat. They have classified each of the pheasant genera as being either solitary or gregarious in social organization, and their mating systems as monogamous (male pairing with 1 female only, per breeding season), harem polygynous (a single male forming a harem containing many females, maintaining their sociality through much of the year), or promiscuous (a.k.a. serial polygyny, establishing individual male display areas that individual reproductive females may visit).

According to scientists, the reasons for such a diversity of reproductive patterns among the pheasants are closely related to their habitat: food availability and habitat structure. In the first suggestion, it is thought that competition for perennial, patchily distributed items of animal prey in tropical forest has resulted in largely insectivorous tropical forest species adopting a solitary lifestyle. Tropical species exhibit asynchronous breeding cycles and serial polygyny. In contrast, open country species tend to feed on seasonal but widespread foods, such as grains and other seeds, along with invertebrate prey that is similarly widespread and hence, are gregarious and harem polygynous.

The second proposal, which relates to habitat structure, centers on the need for male guarding. In open country habitats there is little concealing vegetation, and females can be observed relatively easily by both predators and conspecific males. Therefore, long pair-bonds have resulted, whereby a male guards the female. In open habitats, females of some species remain gregarious throughout the breeding season, so harem polygyny is prevalent, and a single male is able to guard more than one female at a time. In the denser, closed forest habitats, females are more effectively hidden by the dense vegetation, and as a consequence, are not bothered so much by predators and conspecific males. There is, therefore, less need for mate-guarding by the males, and this has led to shorter-lived pair-bonds, such that in the extreme, the pair-bond lasts for the duration of copulation only (McGowan, 1994).

Monogamy: Scientists have considered the genera *Ithaginis, Catreus,* and *Crossoptilon* to be characteristic of scrub and open grassland. All of these exhibit whole-season pair-bonding, a lack of gregariousness among females in the breeding season, and are monogamous. In tropical forest genera, *Afropavo* and *Lophura,* the females are non-gregarious. In *Afropavo,* monogamous pair-bonding occurs which lasts through the breeding season. Because *Afropavo* is considered a generally very primitive form, this is presumably the result of retention of this original trait, rather than a secondarily acquired characteristic. The montane species, white-crested kalij pheasant *Lophura leucomelanos hamiltoni,* also exhibits whole-season pair-bonding. (Yearlings with adult male plumage have been observed feeding with the parent birds in October in Shimla hills, showing that the species reproduces every 2 years, and not every year (author's personal observation)). In the montane forest pheasant genera *Pucrasia, Trogapan,* and *Chrysolophus,* the females are solitary. The western tragopan is monogamous and takes part in nesting, and fledging activities, while in *Chrysolophus,* the pair-bonding lasts only through fertilization (Johnsgard, 1999).

Harem Polygyny: Genera considered as forest-edge or scrub-edge species that forage in the open included *Gallus, Phasianus, Syrmaticus, and Lophophorus.* In all of these genera the females tend to be gregarious and pair-bonding may last as long as the entire season (*Gallus*) or until incubation begins (*Phasianus, Syrmaticus,* and *Lophophorus*) (Johnsgard, 1999).

Fig. 1.29 A sketch of a pair of Lady Amherst's pheasants by T.W. Wood in 'Pheasants: Their Natural History and Practical Management' 2nd edition, 1881, by W.B. Tegetmeier.

Fig. 1.30 A sketch of Reeves's pheasants by T.W. Wood in 'Pheasants: Their Natural History and Practical Management' 2nd edition, 1881, by W.B. Tegetmeier.

Serial Polygyny: Promiscuous pheasants appear to have one of the two mating patterns, which are described in terms of the dispersion of males: promiscuity with aggregated males (lek) and promiscuity with dispersed males. The only pheasants in which the males gather together in groups to display to females are the two Asian peafowl: blue peafowl *Pavo cristatus* and green peafowl *Pavo muticus*. The display dispersion of the blue peafowl, a common sight throughout much of South Asia and in many parks and gardens elsewhere in the world, has been the subject of a number of studies. The clumping of peacocks *Pavo* sp. to display to females falls within the dispersion pattern that has been described as a 'lek', a collection of males that are drawn together for the purpose of attracting females for reproduction. In India, breeding seasonality varies according to the onset of the wet season. Each male establishes a territory in close proximity to other males. These territories are defended by threatening or attacking intruding males, and visiting females are displayed to. Females seem to wander through the various territories on the lek, sometimes singly and sometimes in groups. A study of a feral population of blue peafowl in England has shown that females may visit several different males in a lek, and revisit a preferred male before mating. After mating, the females nest, incubate, and rear the young without the assistance of a male (McGowan, 1994).

On the other hand, promiscuous species in which the males are dispersed rather than clumped together during the breeding season include the Malayan peacock pheasant *Polyplectron malacense*, the great argus *Argusianus argus*, and the crested argus *Rheinardia ocellata*, all inhabitants of rain forests of Southeast Asia. In all of these species the females are nongregarious. Males of all 3 of these species make display scrapes, which are also referred to as 'dancing grounds'; they clear leaf litter from a small area (about half the size of a badminton court) of the forest floor, so that the bare earth below is exposed, and vocalize and perform courtship dance for the female pheasant in the dancing ground. Only adult males are believed to maintain such cleared areas in most years. Great argus females were seen visiting males on their display sites, which are typically widely spaced on small hills in the lowlands; and it has been suggested that females roam through the forest, visiting various males that are in possession of display sites. The male's display is quite extravagant. Following mating, as in the blue peafowl, the female is responsible for the incubation and all the parental care, while males continue to call and maintain their display sites (McGowan, 1994).

MATING SYSTEM ⇒		MONOGAMY		HAREM POLYGYNY	SERIAL POLYGYNY
HABITAT ⇒		FOREST	SCRUB OR OPEN GRASSLAND	SCRUB EDGE OR FOREST EDGE	FOREST
SOCIAL STRUCTURE ⇓	PAIR BONDING ⇓				
FEMALE SOLITARY	WHOLE SEASON	*Afropavo, Lophura, Pucrasia, Tragopan*	*Ithaginis, Catreus, Crossoptilon*	X	X
	UNTIL COPULATION/ INCUBATION	*Chrysolophus*	X	X	*Pavo, Polyplectron, Rheinardia, Argusianus*
FEMALE GREGARIOUS	WHOLE SEASON	X	X	*Gallus*	X
	UNTIL COPULATION/ INCUBATION	X	X	*Phasianus, Syrmaticus, Lophophorus*	X

Table 1.2 - Mating systems and sociality of pheasant genera with respect to their habitat.

Courtship Behavior

THE ARGUS PHEASANT DISPLAYING ITS PLUMAGE

Fig. 1.31 A sketch of argus pheasant males displaying their plumage to females, by T.W. Wood in 'Pheasants: Their Natural History and Practical Management' 2nd edition, 1881, by W.B. Tegetmeier.

It was late afternoon in the heart of the great island of Borneo. I had tramped all day through the jungle and now, at the very end of my trek, had located what I thought was an old dancing arena of an argus pheasant, on a hillock only fifty yards from the bank of the Mujong River. I walked on, located my tiny dug-out, and ridding myself of my jungle-hued clothing, slipped over the gunwale into the dark chocolate waters. In and out of the overhanging roots I swam, every pore of me drinking in the coolness. I clung to a half-submerged vine and let the current sway me back and forth, and searching with my eyes for a chirping insect on an old fallen tree nearby, I suddenly saw, close to my face, a six foot serpent coiled on a branch which still kept its bark above water. I did not recognize it, but it was manifestly a "hot snake" as my Dyak interpreter called poisonous species. I thought how exactly this scene would typify the deadly tropics to my stay-at-home friends. And yet here I was swimming amid the shadows of a strange tropic river, with a venomous serpent watching me, and was probably quite as safe from harm as I would have been in any mill-pond at home (North America). In such a pond a snapping turtle might, in the realm of possibilities, amputate a toe or a foot, or it is far from inconceivable that a rattlesnake or a copperhead might be coiled under my clothing on the bank; I might have to climb out through a mass of poison ivy and in my progress disturb a wasp's nest.

The serpent and I watched each other for a time with respectful interest and then I swam back to my craft and drifted slowly down stream toward the great war-canoe which was my present home. When within a few bends of it I drew myself beneath a maze of branches and lianas and watched the day die over the brown Bornean waters.

The two banks of the river became darker, duskier green, and finally all but the sky-mirrored outer-most leaves changed to black. The sky was pale blue; the muddy water a nameless, beautiful brown. The banks were lifeless most of the day, the jungle folk keeping to the inner forests. Now, however, in the cool of early evening, birds' calls were heard. Small flocks of fruit pigeons dashed over the trees, large mynas perched on tall plum trees, and a family of gibbons shook the branches of a tree in the distance. In a black concavity of the pale, clayey bank a lighter spot appeared, framed by bushes. My glasses (binoculars) showed a wild boar, fore-feet stamping, tushes gnashing and twisted tail flicking. Had he not been against the blackest shadow, he would have been invisible, as he was coated with the mud of the banks. The flies gave him no peace, and he soon turned and climbed awkwardly into the dark jungle behind. The first flying fox of the evening now appeared, flapping slowly and gracefully as a heron; then a score of these giant, five-foot bats came into sight, high in air. As the mynas flew from their trees to some distant roost, the bats swung up to the clusters of fruit and enveloped them like starfishes on oysters, swinging around head downward and eating away with all their might. Distant shrills announced the evening concert of the great five o'clock cicadas.

Then came unannounced, the sight of sights. A few paces to the right of the wild boar's wallow, my eye caught a movement against the water-washed bare face of clay. I pushed my glasses up into focus and there sprang into clear-cut delineation what my eyes had refused to separate from the shadows of the bank – a male argus pheasant drinking from a rain pool a yard or more from the moiled current of the river. It was half crouched, and the motion of the head, alternately raised and lowered, was all that betrayed the bird. The long wings, the gracefully twisted tail-feathers were as motionless as if carved in cameo against the earthen bank. I watched it thus for a minute, two minutes, then my attention wandered for a moment to some creature near at hand, and when I looked back the bird was just disappearing. I had seen my first wild argus, brief though the glimpse had been.

-From *Pheasant Jungles* by William Beebe

Very few bird species can match the drama, the dance, the dazzle, and the intensity of the courtship display of the pheasants, especially that of the blue peafowl, the tragopans, and the argus pheasants. Fortunately for wildlife enthusiasts, the elusive pheasant species have been filmed and documented on websites such as ARKive.org and ibc.lynxeds.com.

As may be expected, there is a positive relationship between the status of a male in the peck-order and his success in mating. Interestingly, the most dominant male may not perform courtship display as frequently as some of the less dominant birds, but he is nevertheless more successful in attracting females. Also, the size of the male's territory has no correlation with the size of his harem; instead, older males are more effective in attracting females than younger males (Johnsgard, 1999).

The major male displays of the domestic fowl, junglefowl, and indeed of most species of pheasants include the following postures (Johnsgard, 1999):

1. Waltzing – It is an important social signal that appears in the context of fighting, when the birds are maturing. It may be performed for a female or an opponent male, not for courtship (which comes later), but for asserting dominance. In its full form, the displaying bird walks sideways around its opponent and holds its back and shoulders in an oblique manner, with side nearer the opponent lower than the other one. Both wings are somewhat laterally extended, but the primaries of the outer wing are lowered to the ground. Waltzing may be virtually stationary with foot movements limited to ground scratching, or it may involve a circular movement around the other bird. Nearly all pheasant species perform a display similar to or equivalent to

waltzing, although it takes greatly differing forms in different genera. Waltzing may be absent in peafowl, *Afropavo*, and some of the peacock pheasants *Polyplectron* sp..

2. Wing-flapping - A highly variable display in which the wings may be moved silently or flapped noisily, including clapping sounds made by the wings striking one another overhead. In many species the display is called "wing-whirring". Most, and perhaps all, male pheasants utter loud advertising calls, sometimes in conjunction with noisy wing flapping displays or other posturing.

3. Tidbitting - The male pecks at the ground or scratches at the ground while giving food calls. In some species of pheasants actual items of food may be picked up and dropped, or may be held in the bill as the food call is uttered.

4. Feather-ruffing - In the domestic fowl the major feathers affected are the hackles of the neck. However, in many other pheasant species the crest, breast feathers, or body feathers in general may also be variably raised or ruffled. In peacock pheasants the entire dorsal plumage is often raised.

5. Head-shaking - In domestic fowl and junglefowl the head is vigorously shaken with circular movements. In other species the intensity or form of head-shaking may vary.

6. Tail-tilting, tail-spreading, or tail-wagging - Tail exhibition (or the exhibition of specialized tail-coverts, as in Peafowl) are common forms of visual signaling in pheasants and usually occur in combination with waltzing, side display, or frontal display. The mating display of the blue peafowl involves the following. The male's iridescent train is erected by the cocking of the rather long but nondecorative rectrices. The lateral tail-coverts extend horizontally and even downward, so that they hide the wings, which partially droop. The whole body is inclined somewhat forward. The head and neck are held erect, in the middle of the radiating pattern of ocelli formed by the tail-coverts. When thus facing a female, the male suddenly performs a quivering shake,

which causes the iridescent train to shimmer and the feathers of the wings to rustle audibly. As female approaches, the male may actually turn away from her, causing the hen to move around and face him. This is repeated several times before the female crouches. The male then rushes towards her in a characteristic hoot-dash posture. During mating the train is lowered and mounting occurs in the usual galliform manner.

Fig. 1.32 A sketch of golden pheasants by T.W. Wood in 'Pheasants: Their Natural History and Practical Management' 2nd edition, 1881, by W.B. Tegetmeier.

7. Bilateral wing-lowering - Although not well developed in junglefowl, this is a major display in many pheasants and reaches its peak in peafowl and peacock pheasants. In peafowl, it has completely replaced lateral display posturing.

8. Wattle, comb, or facial engorgement - Nearly all pheasants utilize the temporarily enlarged areas of facial skin as sexual or aggressive signals. The Bulwer's pheasant (a.k.a. wattled pheasant) represents the culmination of this trend, whereas in others such as blood pheasants it is hardly noticeable.

9. Cornering - This display is evidently an important precopulatory display in the genus *Gallus*. In this behavior a male runs to a corner of its enclosure, stamps its feet, and sits down, making a low purring sound. It serves as a nesting invitation display and perhaps helps the female decide on a nest-site location.

10. Crouching - This female pheasant display is essentially an accepting posture. In all pheasant species it takes a similar form.

Fig. 1.33 Blue peafowl male displaying to peahen in the dry summers of Ranthambhore National Park, Rajasthan, India. Photo by Aditya Singh, June 2010.

11. Other species of pheasants performs certain additional displays. For example, male tragopans have elaborate display postures associated with exhibition of their normally hidden gular lappets (brightly coloured fleshy membrane on the throat) and 'horns' (two fleshy horn-like structures on the head that are visible during courtship display).

Fig. 1.34 A. A painting of the 5 species of tragopans, showing distended gular lappets, by Henrik Gronvold in 'A Monograph of the Pheasants' (created 1918-22) by William Beebe. B. A male Satyr tragopan displaying by distending its gular lappet, and erecting its horns, in a breeding center of Galliformes in Braga, Portugal. Photo by Hugo Barbosa.

Nesting

No actual proof of the nesting in trees of these gorgeous pheasants had been forthcoming, until I was fortunate enough to stumble upon circumstantial evidence of so positive a nature that I include it in detail. In this account, the tragopan entered upon the scene wholly unexpectedly.

In this range of native Garhwal, I had set up my observation tent on a sloping hillside of pine. I placed it under and within the dense sweeping branches of a young deodar, so that it made a formless mound of green, indistinguishable from the mass of dark needle foliage about it. Here I left it for three days, and then entered it one morning with the intention of observing more closely some cheer pheasants, which were accustomed to pass over this slope twice a day. An hour after I had begun my vigil, I cut a new observation slit in the rear, for the purpose of finding the author of a sweet, silvery thread of warbling notes. A moment before, they had been uttered within a foot of the tent, and now I found the bird had flown to the short depressed branches of a silver spruce, forty feet up, and not far away. The activity of the little bird, whatever it was, prevented my identifying it; but in searching for it, I discovered a rough mass of sticks, lodged close against the trunk, and partly overhung and concealed by several of the silvery-green needle-fans of this splendid conifer. I marked it down as an object for examination when I should leave the tent, and, after the usual few minutes of exercise and massage within my little green mound, I returned to my survey of the hillside.

Passing over many unimportant but interesting bits of forest life which I observed on this memorable day, I at last caught a low, pheasant-like chuckle, which made every nerve tingle like an electric shock. I peered out, but could see no sign of life. Then the chuckle again and the quaver of needles, and on the branch below the stick nest, I saw a large bird. Even then, tragopans were so far from my mind that I stared in unrecognizing bewilderment. Once more the low gurgling chuckle came, and the bird walked unsteadily to the trunk and leaped up to the nest itself. Then I realized that I was looking at a western tragopan hen, and a few days later, I was to hear the self-same low chuckle given as the vesper song of a cock bird.

The tragopan, in her dull, mottled garb, was almost invisible as she stood motionless beside the nest in the shadow of the spruce foliage. Soon she began to crane her head and neck about, and, bending low down, busied herself in some way invisible from where I watched. At last she jumped down to the branch below, then to the next, and so on, making a complete circuit of the trunk as she descended, and finally, when out of my sight, flew with but a low rustle of wings to the ground. For forty minutes, I saw or heard nothing more, and then the crackle of a twig set me on the *qui vive*, and I soon saw her near the nest. Again she wrought silently at the nest, and again descended her resinous stairway. Once more she returned, this time with a beakful of leaves, which I could distinctly see, as they were so unlike the needles through which she ascended. She soon went away as before, and I never saw her again, although I waited until late afternoon, when my abused body would allow no further insult, and for very agony I had to leave my shelter and roll about upon the turf outside. Once having thrown caution to the winds, I climbed the tree with some difficulty. Knowing that the wary bird would easily detect my clumsy trail of sap-bleeding footprints, I tied together the entire nest, brought it down, and made a careful analysis of the structure. A glance showed that it was not the work of the pheasant, but an old nest of some other bird; this disappointing fact being only too evident from the weather-worn character of the well-woven substructure of sticks and bleached grass. The lining was as obviously of very recent date; indeed, the green leaves of oak and some unknown weeds were still almost fresh and unwilted, while the twigs – a dozen or more with a strong aromatic scent, were still sappy at the end, for all had been freshly broken off, and none were dead or dried. All had been plucked within 48 hours. This was the work of the tragopan which I had been watching, although she could not have brought all the twigs and leaves on the 3 trips under my observation. She must have begun work on the previous day.

I now turned my attention to the nest proper. I found that the twigs and grass were not nearly so bleached as the old dried stems about me on the ground, and traces of green near the nodes of the grasses seemed to make certain that it was this year's nest. Two small fragments of (egg) shell, which had sifted down into the matted lining, might have been parts of the egg of a crow or raven, or of many other species. The general character of the nest was corvine – no more could be said.

One can readily see how many terrestrial dangers would be avoided by birds of this size nesting in trees; but, on the other hand, if they are in the habit of utilizing the large stick nests built and already used by other birds, they are running considerably more risk than if they built a nest themselves. This nest was remarkably well concealed. But most used nests are much more in view. Constant approach of the parent birds breaks or wears away the adjacent leaves and twigs, and by the time the brood of the rightful owners is ready to leave, the nest is far from well concealed. Also, fearless and pugnacious corvine birds usually build in plain view, a site which would give a hen tragopan but short shrift. The greatest danger attendant on arboreal nesting would be the Himalayan langur monkeys, and the betrayal to eagles by inquisitive jays and crows.

- From *Pheasant Jungles* by William Beebe

Birding enthusiasts also delight in noting how each species of bird makes a particular type of nest, differing from other species in its shape, size, and material composition. For example, the nest of a red-vented bulbul is a grass bowl 3-4 inches in diameter, while that of a painted stork is a platform of juxtaposed twigs, about 1 meter in diameter.

Most of the pheasant species are ground-nesters i.e. they build their nests on the ground, usually at well hidden locations such as under thickets or low overhanging branches of trees. In case of the Indian blue peafowl *Pavo cristatus*, the nesting season is quite varied, and is apparently related to the timing of the wet monsoon season. In southern India and Sri Lanka, nesting occurs from January to April. Along the foothills of the Himalayas, nesting also may occur as early as March or April, but usually begins with the start of the summer monsoon, or about the middle of June. Nests of the blue peafowl are typically well concealed and often are located under thorny shrubbery such as *Lantana* or *Zizyphus*. In some low areas, where flooding often occurs, it may nest in elevated sites, such as the forking trunk of a *Ficus* tree. The clutch size has been reported to range from 3-6 eggs. Incubation is done by the female alone and lasts for 28-30 days. Nest-site selection and associated egg-laying behavior has been studied a little in domestic fowl. It has been learned that females typically visit several potential nest sites before selecting one, and that young females appear to be very nervous about choosing a site. Concealed sites are more attractive for nest-building than open sites (Johnsgard, 1999).

Fig. 1.35 A. A nest and B. Newly hatched chicks, of the ring-necked pheasant in Oregon, USA.

Nest-building activity in all Galliformes, including pheasants, takes a similar form. Female pheasants do not make elaborate nests. Nearly all pheasants construct simple nests in shallow scrapes, sometimes lined with grass, ferns, or feathers. However, in tragopans *Tragopan* sp., crested argus *Rheinardia ocellata*, and great argus *Argusianus argus* the nests are often placed in elevated situations i.e. in trees or dense cane thickets. In the equatorial lowland forest of Southeast Asia, forest-dwelling pheasants seem to prefer sites at the base of some prominent vegetation features. For example, the crestless fireback pheasant

Lophura erythrophthalma gathers twigs or similar material together to build its nest between buttress roots of a tree (Johnsgard, 1999, and McGowan, 1994).

Clutch Size

Once a nest-site has been chosen, the nest is constructed and the eggs are laid. The average numbers of eggs per nest, i.e. the clutch size, varies greatly between species, with relatively large clutches being typical of grain-eating, edge or open-country species such as *Phasianus* and *Gallus*, and small clutches, of as few as only 1 or 2 eggs in forest-dwellers such as *Polyplectron* and *Argusianus*. The Malaysian peacock pheasant is the only phasianid that is known to lay a single-egg clutch. The tropical forest-dwelling gallo-pheasant species (genus *Lophura*) also lay smaller clutches than the northern species; the tropical species crestless fireback *Lophura erythrophthalma*, for example, lays 4-5 eggs, whereas the temperate species silver pheasant *Lophura nycthemera* and kalij pheasant *Lophura leucomelanos* lay 6-12 or even up to 15 eggs. In seasonal temperate regions, the hatching of phasianid chicks tends to coincide with annual peaks in food supply (seeds, grain, etc), but such peaks do not occur in tropical forest. Therefore, the larger clutches in temperate grasslands may be a consequence of dependable food sources being available (McGowan, 1994). In both types of habitats, only a percentage of the eggs actually hatch and survive up to adulthood, preyed upon by ravens, weasels, snakes, cats, etc.

Incubation is usually carried out by the female only. In monogamous pairing, males are concerned with territorial defense during this time, and in non-monogamous pairings, males may attempt to attract other females with which to mate (McGowan, 1994). The durations of incubation among the pheasants range from 18 to 29 days, with the longer duration typical of such genera as *Argusianus, Pavo, Lophophorus*, and *Crossoptilon*, whereas duration of less than 23 days occur in forms such as *Pucrasia, Chrysolophus*, and *Polyplectron*. After hatching, the eggshells are left in the nest, and the young chicks are taken away from the nest site shortly after (Johnsgard, 1999).

Fig. 1.36 A sketch of a pair of silver pheasants by T.W. Wood in 'Pheasants: Their Natural History and Practical Management' 2nd edition, 1881, by W.B. Tegetmeier.

In a few species (*Tragopan, Rheinartia*) the young attain a limited flying ability when only a few days old, whereas in others, the flight feather growth is relatively slower and fledgings may require a week or more. In most of the species, adult male plumage and sexual maturity is attained in the 1st year, but in some genera adult plumage and full breeding capabilities are attained after 2 (eg. *Tragopan, Lophura*), or even 3 years (eg. *Pavo, Rheinartia, Argusianus*) (Johnsgard, 1999).

Chapter 2

A Historical Perspective

The study of any branch of science (or any subject for that matter), is incomplete without learning about the historical places, events, and people associated with it. Wildlife has always fascinated mankind, and it has shaped human cultures to a great extent. The time period from 15th to 19th centuries has been the era of explorations and species discoveries that has resulted in making available food plants, medicinal plants, wood, fibers, etc. from different continents, to all in the global marketplace. Since the start of scientific taxonomy in the 18th century, the flora and fauna of the continents have been classified into various taxa, and this has helped in their identification and conservation. The world witnessed a peak in environmental degradation in the 20th century, as well as the emergence of conservation science. Today, scientists are engaged in wildlife research to assist the conservation programs in countries around the world; and awareness about natural history and conservation science are an essential tool to achieve sustainability of economies dependent on natural resources.

Fig. 2 A painting of a male golden pheasant from *A Hand-book to the Game-birds* Vol. I and II, by W.R. Ogilvie-Grant, 1897.

Chapter 2

A Historical Perspective

Pheasants in Human Lives

All the species of order Galliformes have been closely associated with man since time immemorial, and are unrivalled among birds for their impact on man. This is primarily because, as large birds, they make a protein-rich meal, and provide nutritious eggs. Among these species, it is perhaps the pheasants which have had the deepest impact on man - including religious significance in many parts of Asia, providing food and sport, and even sustaining local economies in developed countries of the world (McGowan, 1994).

Many pheasants, francolins, and partridges have been providing a varied diet for the peoples of parts of China, South Asia, Southeast Asia, and Africa. The hunting of these birds as food sources has depended, to an extent, on the religious beliefs prevalent in the regions. For example, the many species of francolins, quails, and guineafowl found in the Horn of Africa, benefit from the non-hunting beliefs of the local Christians, and of Muslims; and the ground-dwelling forest species of Bhutan in the east Himalayas are still abundant thanks to the Buddhist tradition of that country (McGowan, 1994).

Some of the galliform species, hardy enough to stay close to human settlements and provide a food source, would also be taken along by travelers and explorers, and survived long journeys across land or aboard ships. The females would produce regular food in the form of eggs. So, some of the birds were well known outside their native ranges long ago, and they have become important to man wherever they have been introduced (McGowan, 1994).

Some pheasants were well known over 3000 years ago, when they were depicted in embroideries and paintings in China and elsewhere (McGowan, 1994). The pheasants are also mentioned in a collection of poems known as 'The Odes', which form a cornerstone in Chinese literature. It is a selection of more than 300 poems from as early as 1000 BC compiled by the great philosopher Confucius. The poems are from the region in the Yellow river valley in China, and have several references to plants, animals, and birds. The poems paint a picture of everyday life in the river valley – peasant women expressing joys of farming and harvest; mention of trees of cherry, papaya, peach, plum, pear, and mulberry; crops of wheat, millet, mustard, pepper, hemp, and liquorice; lotuses and bamboo; animals and birds such as fox, goose, quails, magpies, egrets, and pheasants – nature's bounty, peppered with human emotions of love, joy, anxiety, sorrow, and including the life of court officials, of conduct and discipline. One such poem, titled 'Xiong Zhi' (Anonymous, 1998), describes a woman upset at her mate for disturbing the pheasant she was observing –

The male pheasant flies away,
Lazily moving his wings.
The man of my heart! --
He has brought on us this separation.

The pheasant has flown away,
But from afar comes his voice.
Ah! the princely man! --
He afflicts my heart.

Look at that sun and moon!
Long, long do I think.
The way is distant;
How can he come to me?

All ye princely men,
Know ye not his virtuous conduct?
He hates none; he covets nothing; --
What does he which is not good?

Similarly, Indian Sanskrit writer Kalidasa's romantic poems are peppered with the names of plants and animals of the Indian plains, described in two of his works, *Rtusamharam* (The Gathering of the Seasons) and *Meghadutam* (The Cloud Messenger). Kalidasa, whose work is from around the 1st century BC, is considered the foremost Sanskrit writer. Some examples from *Rtusamharam* (Rajan, 1989) –

'Rains'

Implored by chatakas (cockoos) tormented by great thirst,
and hanging low weighted down by large loads of water,
massed clouds advance slowly, pouring many-streamed rain:
and the sound of their thunder is sweet to the ear.

A bevy of peacocks that sound ever-delightful,
eagerly watching out for this festive moment,
caught up in a flurry of billing and preening,
now begin to dance, gorgeous plumage spread out.

And, 'Autumn'

The earth is bright with Kasa (a tall grass) blossoms,
nights with the cool rays of the moon;
streams are lively with flocks of wild geese
and pools are filled with lotuses;
groves are lovely with flower-laden trees
and gardens white with scented jasmines.

Lost is Indra's bow (the rainbow) in the folds of the clouds;
lightning, the sky's banner, quivers no more;
egrets no longer beat the air with their wings;
peacocks do not watch the sky with upturned faces.

Ancient Greeks, like the play-write Aeschylus (524-456 BC, was a Greek tragedian who wrote plays such as 'Oresteia' (a trilogy), 'The Suppliants', 'Seven Against Thebes', and 'The Persians'), knew of the ring-necked pheasant *Phasianus colchicus* of Colchis (an ancient kingdom on the east coast of the Black Sea, now primarily Georgia, just north of Turkey); and the Indian peafowl *Pavo cristatus* featured in Greek mythology (McGowan, 1994). Pheasants also find ample reference in the Japanese style of poetry, called 'Haiku', which evolved in the 17th century. The Tibetan art of 'thangka' painting often depicts species of montane birds and animals, and pheasants are no exception.

During the Mughal era in India, especially during Jahangir's rein, the wildlife of India was depicted in paintings, created by artist Ustad Mansur, and his apprentices, who were employed by the emperor. Emperor Jahangir was a great observer and admirer of wildlife, and several foreign visitors brought him exotic animals as gifts. These were also painted by Mansur with exceptional accuracy. Towards the end of the Mughal era, in the late 18th century, the paintings and literature originating from the courts of Emperors Babur, Akbar, and Jahangir, were plundered (Ali, 1979). As a result, most of them are now found in museums around the world, such as the Smithsonian Museum, the British Library, Victoria and Albert Museum, Harmitage Museum, and museums in Iran (see 'Further Reading' for Smithsonian Museum link).

Red Junglefowl

Among pheasants, the species which has had the most impact on human life is the red junglefowl *Gallus gallus*. There is strong evidence to suggest that the widespread red junglefowl is the progenitor of the domestic chicken - an important part of human life everywhere. The domestication of this species has featured prominently in human history. The domestic fowl has provided a living and reproducing source of food for seafarers and colonizers. It has also stirred a wide range of fears, hopes, and emotions; it has been invested with religious symbolism, and has provided sport and recreation of various kinds (McGowan, 1994).

Fig 2.1 Photo of a male red junglefowl in Kaziranga National Park Assam, India, by Aditya Singh, March 2009.

Indeed, the extent to which man has been associated with the domestic fowl is indicated by the influence that it has had on languages of the world. In English, for example, there are many words or phrases referring to the strength, aggressiveness, and strutting nature of the cock fowl ("cock", "cocky", "cock-a-hoop", "cock-sure", and many other phrases). Traditional Malay sayings make reference to pheasants such as junglefowl and argus (McGowan, 1994). There are also several Hindi proverbs and aphorisms using the junglefowl and blue peafowl as metaphors.

Fig. 2.2 A brass peacock lamp from south India in a handicrafts emporium in New Delhi. Photo by Anita Chauhan.

There is uncertainty over when and where the wild red junglefowl was first domesticated. It seems that the junglefowl was kept by the civilizations that inhabited the Indus Valley in the 3rd millennium BC (i.e.

2600 BCE, Harappa and Mohanjodaro civilisations), and India is most probably the original centre of its domestication (McGowan, 1994).

Piecing together what is known about the journey of the domesticated chicken through both time and space, tells a revealing story about the history of mankind's travels and explorations, as chickens, once identified as a source of food, accompanied these voyages everywhere. Not long after the domestication of the fowl in the Indus Valley, it appeared in Persia, and spread through Crete and Phoenicia (which is now primarily coastal Syria), to central and north-western Europe, where it had arrived by 1500 BC. Further south, the keeping of domesticated chickens spread through the western part of the Mediterranean from Greece. Domestic fowl were present in both Egypt and China by 1500-1400 BC, although they were possibly not commonly kept in the former until several centuries later (McGowan, 1994).

It seems that chickens were kept by people throughout many parts of the world, by the 1st century AD. Red junglefowl were already on the South Pacific islands by the time European settlers first visited them. It is believed that early Pacific seafarers took chickens with them on voyages throughout the Micronesian, Melanesian, and Polynesian islands. Some of the chickens became feral on most of the islands that were visited (McGowan, 1994).

They were found on the Mariana Islands (an arc shaped archipelago consisting of 15 volcanic mountains in the Pacific ocean, lying north of New Guinea, named after Spanish Queen Mariana in the 17th century) at the time of Magellan's voyage in 1521 [Ferdinand Magellan was a Portuguese explorer who served King Charles I of Spain in search of a westward route to the 'Spice Islands' (today's Maluku Islands of Indonesia). The Strait of Magellan at the southern tip of S. America, and the Magellanic penguins are named after him], and were common in the wild on the coral islands of Tinian in the south of Mariana chain by the time Pascoe visited it in 1742. Explorers such as James Cook (British explorer, navigator and cartographer in the Royal Navy), and William Bligh (an officer of the British Royal Navy and a colonial administrator) reported that they were already found throughout the Pacific islands. Now, there are red jungle fowl living wild on many of these islands. Chickens were taken to the Americas by Europeans - Christopher Columbus (1451 – 1506, was an Italian explorer, colonizer, and navigator) taking these birds with him on his 2nd voyage to the West Indies in 1493 (McGowan, 1994).

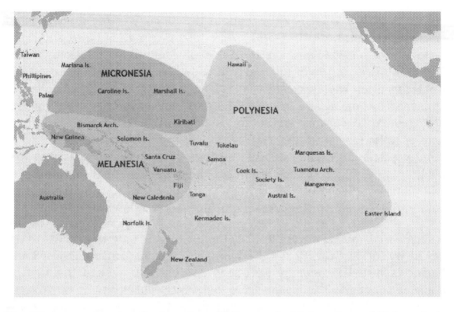

Fig. 2.3 A map showing the location of Micronesia, Melanesia, and Polynesia in the Pacific Ocean.

Not only have domestic fowl been taken to more parts of the planet than virtually any other animal, but they have also entered into a great many aspects of human life, invoked as guardians of spiritual and physical well-being. Many medicinal benefits have been attributed to the cock at various times in history, and in various parts of the world. The cock has also been of great religious importance throughout its association with mankind. To the Zoroastrians of ancient Persia, who were known as fire worshippers, the cock was an object of sacrifice. The Greeks dedicated the cock to three Gods: to Apollo, as it heralded the rising of the sun; to Hermes because its crowing called man to work; and to Aesculapius, because of the virtues of retiring to sleep and rising early. The domestic chicken is believed to have been introduced to Palestine by about 200 BC. Both the New Testament and the Talmud have references to it. Traditionally, the cock has been the sacrificial animal of the Jews, being an atonement offering at Yom Kippur, a man giving a cock and a woman a hen. The cock was used to ornament Gothic churches as it was the standard of war to the Goths (McGowan, 1994).

The domestic chicken is used in modern medicine by scientists to understand some of the formative processes of vision, learning, and memory (McGowan, 1994). Also, the red junglefowl genome has been mapped completely, and geneticists have found useful 'control sequences' of genes, which are of importance to the field of medicine.

The Chinese have a very close relationship with many species of pheasants. The brown eared-pheasant *Crossoptilon mantchuricum* is considered a symbol of bravery because of the battles that the males fight. The Generals of most emperors from the times of the Warring States, in 475-221 BC, to the Qing Dynasty, AD 1644-1911, placed the tail feathers of this pheasant in their helmets. The rooster, the descendant of the red junglefowl, is the only bird included among the 12 animals of the Chinese Zodiac, in which each animal represents a Chinese year in a 12 year cycle. The rooster symbolizes the 5 virtues of being a good omen, authoritative, courageous, good, and confident. It is a good omen because of the bearing with which it carries itself; its comb lends it authority, and the spurs show its courage. It is good because it always shares its food, and the assurance with which it greets the sunrise indicates its confidence. To celebrate the last 'Year of the Rooster', 23rd January, 1993 to 9th February 1994 in the Gregorian calendar, the Beijing Natural History Museum held an exhibition to show how important this species, and the other galliforms, have been to man, and to highlight the plight of the Chinese pheasants in particular (McGowan, 1994).

Blue Peafowl

By and large not used by man as a staple source of food or sport, the Indian peafowl (a.k.a. blue peafowl) *Pavo cristatus* has mostly been admired for the beauty of the male. Reference to the Indian peafowl does not seem to go as far back, as that to the red junglefowl. Although it is constantly referred to in Indian mythology and Sanskrit writings, the first mention of the bird outside India was not until the time of Solomon (King of Israel, 1011 BC – 931 BC). His Phoenicians (ancient civilization near Mediterranean Sea), voyaging to the coast of India, brought the Indian peafowl back to the Syrian mainland, and also to the Egyptian Pharoahs (McGowan, 1994).

Fig. 2.4 A male blue peafowl perched on a mango tree, New Delhi, India. Photo by Anita Chauhan, 2010.

The species was not widespread in Greece, until Alexander the Great brought it back from his travels in the latter part of the 4th century. Alexander was so struck by the bird's beauty, that he imposed a severe penalty on anyone caught harming it (McGowan, 1994).

In its Indian homeland, the peacock is sacred to the Hindus. There is considerable folklore associated with the peafowl across the length and breadth of India. It is considered one of the main destroyers of snakes (McGowan, 1994). The Hindu gods, Krishna and Ganesha, are always depicted wearing a peacock feather on their head. The goddess of knowledge, Saraswati, is always depicted with a peacock. 'In Sri Lanka, Singhalese medicine system uses the fat of the bird as a cure for rheumatism, sprains, and dislocations' (McGowan, 1994).

Outside its native range, the fact that the peacock has been considered a bird of great beauty and magnificence has led to its symbolizing wealth and power. For example, during Roman times a breeder of peafowl could make a good living by providing birds for lavish feasts. In England, in the Middle Ages, barons served up roast peacock at their banquets to prove their wealth (McGowan, 1994).

Ring-necked Pheasant

Apart from using the Indian peafowl as a symbol of their wealth, members of the British aristocracy, in the Middle Ages, demonstrated their status by hunting other phasianids, most notably, the ring-necked pheasant *Phasianus colchicus*. Native to the eastern Palearctic, where it is distributed from the eastern shores of China west to the Caucasus Mountains, it is commonly called simply 'the Pheasant', or the 'common pheasant' in most of its introduced range, as it is the only member of the subfamily to occur in regions outside Asia (McGowan, 1994).

The ring-necked pheasant is alleged to have reached Europe, when Jason [ancient Greek mythological hero who sailed on the ship 'Argo' to procure fleece of the gold-haired ram from Colchis (Georgia)] brought it back on the 'Argo' after his search for the Golden Fleece. Indeed, both words *Phasianus* and 'pheasant' come from the Greek word *phāsiānos*, meaning '(bird) of the Phasis'. Phasis is the ancient name of the main river of western Georgia, currently called the Rioni. But, it has also been suggested that the Romans were the first to bring pheasants to this region. The Romans may have introduced pheasants to Britain, Italy, Germany, and France, from where the record of ancient recipes indicates that they were avidly sought after. Records indicate that ring-necked pheasants were introduced in the UK in the 11th century and in the USA in the 18th century. Within the last 300 years, the ring-necked pheasant has been introduced in most of Europe, North America, some Australian islands, and New Zealand, among other areas (McGowan, 1994).

Whether the species was introduced into Britain by the Romans or much later during the Norman Conquest, it was certainly established and breeding in the wild by the late 15th century. It was granted some degree of protection by the royal courts. Charles ll, for example, only allowed the richest 5% of landowners to hunt, and imposed harsh penalties on lawbreakers (McGowan, 1994).

After the establishment of the pheasant as a game bird in England, the prestige associated with the sport of game-shooting resulted in protection of forests in which the pheasant thrives. It has been suggested that the revenue earned through the popularity of pheasant-shooting has played a major role in the preservation of lowland woodlands in southern England. The large private forest estates, under economic pressure, would have otherwise converted all the land to agriculture (McGowan, 1994).

Fig. 2.5 A. A male common/ring-necked pheasant. B. A male green/Japanese pheasant (the subspecies of common pheasant).

The money spent on managing British ring-necked pheasants reflects their importance to some rural economies. The British government's Conservation Department spends a considerable percentage of resources for managing the game-birds (including the ring-necked pheasant) and their habitat, for the purpose of the sport of shooting- a recreational activity that brings in millions of pounds of revenue and provides a boost to their rural economy (McGowan, 1994).

Fig. 2.6 A. Painting titled – 'The Last of the Flush', and B. Painting titled – 'The Count' - from 'The Pheasant' in Fur and Feather Series, 1895, by H.A. Macpherson, AJ Stuart-Wortley, AI Shand. Illustrations by A. Thorburn. C. Painting titled – 'TOHO!' – from 'American Partridge and Pheasant Shooting', 1877, by Frank Schley.

In the USA, pheasant-shooting is a popular recreational activity. Millions of captive bred ring-necked pheasants from pheasant-farms are released into hunting reserves, where they are shot for sport. Millions of dollars worth of revenue is generated from pheasant-hunts every year. Field studies in the US (for example, in the states of Iowa and Illinois), have revealed that the practice of setting aside a part of private farmlands (called farm set-aside), and public wilderness areas, for the sport of pheasant hunting, has resulted in the conservation of many native plant and animal species. The government (US Department of Agriculture), through its 'Conservation Reserve Program', has been providing economic incentives to private farmland owners, to set aside a part of their farmland for wildlife conservation, and for improving water quality, and prevention of soil erosion. In recent years though, rising commodity prices and land prices have caused a decline in the habitat suitable for pheasants, and as result, the number of people engaging in this recreational sport has also decreased (USDA, 2013 and Eligon, 2012).

Pheasants as Aviary Birds

In addition to the importance of phasianids as game birds, some species have also been used as ornamental birds due to their spectacular beauty. Ornamental pheasants have been sought so that they may be exhibited in both public zoos and private aviaries, or to wander freely across the lawns of large houses, as vivid adornments. The extravagant plumes of some species are of cultural importance in several countries. For example, the magnificent tail feathers of Reeves's pheasant *Syrmaticus reevesii*, which may reach 150 cm in length, have traditionally been used to adorn headdresses in the famous Peking Opera of China. Nowadays, plastic imitations are used, which is a blessing for the species that now has a much reduced and fragmented population in the wild. Similarly, the train feathers of the green peafowl *Pavo muticus* were, until in recent times, used in Javanese dances (McGowan, 1994). Feathers of pheasants, such as ring-necked pheasant and blue peafowl (and other gallinaceous birds), are used in bridal headgear, and in 'Fascinators' and other hat decorations, by Christian women.

Fig 2.7 A painting of pheasants in an aviary.

Fig. 2.8 An advertisement for an aviary manufacturer, from at the back of *'Pheasants: Their Natural History and Practical Management'* 2nd edition, 1881, by WB Tegetmeier.

The leisurely appreciation of these fine birds appears to have been of western European origin, as the aristocrats of several countries employed collectors and dealers to bring back examples of the many dazzling animals that were to be found in distant lands that, at that time, were the colonies of European countries. The aristocrats who maintained collections during the late 19th century in Paris and London, were continually sent new pheasants. Indeed, the 1860's and 1870's must have been especially exciting times, as many species new to science were dispatched live to Europe, to be displayed before the public and admired (McGowan, 1994).

Apart from the red junglefowl, ring-necked pheasant, and Indian peafowl, the first pheasant recorded alive in Europe was probably the golden pheasant *Chrysolophus pictus*, kept as early as 1740, when Eleazar Albin (English naturalist and illustrator) reported in his 'Natural History of Birds', that several of them were in the collections of "our nobility and some curious gentlemen". The grey peacock-pheasant *Polyplectron bicalcaratum* soon followed, judging by the paintings of pheasants present in one Mr. John Munro's collection in London during 1745. Subsequently, Reeves's pheasant *Syrmaticus reevesii* was kept in Macao, China by an Englishman by the name of Beale in 1808. Reeves (1774-1856, was an English naturalist. He was appointed Inspector of Tea for the British East India Company in 1808) himself bringing one of the cocks to Europe in 1831, and describing the species. The green peafowl *Pavo muticus* appeared in aviaries in Cape Town, South Africa, in 1813, sent from Macao, and reached Europe in 1831, by which time Lady Amherst's pheasant *Chrysolophus amherstiae* had also arrived. This latter species only survived for a short time in London Zoo, having been kept in India (in tropical conditions) for 2 years prior to shipment to England. In the mid-1820's, the King of Ava (a kingdom in what is now upper Myanmar), presented two males from the mountains of Cochinchina (a French colony in South Vietnam in 1862-1954) to Sir Archibald Campbell (an officer in the British Army), who passed them on to the Countess Amherst (wife of William Pitt Amherst, Governor-General of India between1823-1828) (McGowan, 1994).

These occasional importations were nothing compared to the influx of new species that took place during the second half of the 19th century (McGowan, 1994).

The keeping of pheasants, and to a lesser extend partridges, quails, francolins, and snowcocks, has become increasingly widespread since those early days, when it was the exclusive preserve of wealthy aristocrats, naturalists, and the like. Virtually all species of pheasants, and many perdicines, are now maintained in captivity in many countries. The appeal of this family to such a great number of people has resulted in many species being widely bred. Indeed, there are far greater numbers of most of these species in the collections of private aviculturalists, than there are in public zoos and bird gardens. It is increasingly realized that these captive collections must be properly managed if they are to contribute to the long-term conservation of the members of this family (McGowan, 1994).

Previous Works on Pheasants

Although Indian ornithology has its roots spanning the ancient Hindu literature and Mughal literature, the scientific documentation of birds as it is known today was accomplished by the British colonists, in the 19th and early 20th century.

In the British colonial times in India, hunting of game-birds was a sport popular among Indian maharajas (some game being taken by the common natives, though the population was largely vegetarian). The British also found the abundant jungles containing plentiful game in India to be the best hunting grounds in Asia.

Much of the ornithological observations published in newspapers, journals, and books in colonial India (and in Britain in this period), were made by the British army and government officers posted in India. They spent their spare time observing wild birds and animals, and some of them- like Mr. A.O. Hume and Mr. Stuart Baker- were also hunting sportsmen who recorded the habits and distributions of the game-birds found on their expeditions. The first comprehensive book was written by Dr. T.C. Jerdon – *The Birds of India*, in 1862. Prior to that, ornithologists like Edward Blyth (curator of the Asiatic Society of Bengal's museum), Colonel S.R. Tickell, and B.H. Hodgson were known for their journal articles and books, and for discovering previously undescribed species of birds and mammals.

Typically, the writers would take the help of curators in the British Museum (and in Indian natural history museums), and consult university departments, and then publish their book in India, or from London. They also engaged artists, who made accurate paintings of the birds based on the museum specimens, to accompany the text. So, it is remarkable that many of the pioneering books on Indian ornithology in British India were written, not by professional biologists, but by Civil Servants with an interest in hunting and bird-watching. These books were written for the hunting sportsmen and amateur naturalists who would have liked to identify the specimens they had observed, and gain knowledge about their distribution and habits.

Fig. 2.9 Photos of A. the Natural History Museum, London, UK, B. the Indian Museum in Kolkata, India.

Fig. 2.10 A. Dr. T.C. Jerdon, and B. Mr. A.O. Hume. Portraits from '*The Nests and Eggs of Indian Birds*' Vol. 1, 1889, by A.O. Hume.

Table 2.1 It is interesting to read about some of the pioneering writers, and how ornithological documentation in India developed over time.

Year(s) of Publishing	Name of the Book	Authors
1862-64	'The Birds of India' (in 3 volumes)	Dr. T.C. Jerdon He was a physician and a biologist, a Surgeon Major in the Madras Army when he published the book. He was first employed by the East India Company, and spent more than 25 years in India. He was a regular contributor to ornithological journals. Several bird species, such as the Jerdon's Courser, carry his name.
1879-81	'The Game Birds of India,	Mr. A.O. Hume and C.H.T. Marshall

	Burmah and Ceylon' (in 3 volumes)	A.O. Hume was a civil servant, political reformer, and amateur ornithologist in British India. He was one of the founders of the Indian National Congress, a political party that was later to lead the Indian independence movement. Hume has been called "the Father of Indian Ornithology" and, by those who found him dogmatic, "the Pope of Indian ornithology", as he was very picky in choosing articles for publishing in a quarterly journal that he started and edited, called *Stray Feathers*, published between 1872-1899. He built a network of ornithologists reporting from various parts of India, including – T.C. Jerdon, C.H.T. Marshall, E. Oates, W.T. Blanford. He had a vast bird collection, which he later donated to the British Museum (now Natural History Museum). The Hume collection, as it went to the British museum in 1884, consisted of 82,000 specimens of which 75,577 were finally placed in the Museum. C.H.T. Marshall was a British Army Officer, serving in Punjab, India. In his spare time, he collected birds in Punjab and the Himalayas, and sent these to A.O. Hume. He was the brother of G.F.L. Marshall, with whom he published ornithological articles in *The Ibis* (the journal of British Ornithologists' Union (1859 to present)).

1889-98	'Fauna of British India including Ceylon and Burma: Birds' (in 4 volumes)	Mr. E. W. Oates and Dr. W.T. Blanford The first two volumes were written by Oates, and the next two by Blanford, all four were edited by Blanford.
	And, also by EW Oates - 'A Manual of the Game Birds of India' (1898-99, in 2 volumes) and 'A Handbook to the Birds of British Burma'. Edited the 2nd edition of Hume's 'Nest and Eggs of Indian Birds'.	Mr. Oates was a civil servant, and spent 32 years in the Public Works Department of India, devoting his spare time to the ornithology of British India. He consulted the British Museum (Natural History) for writing his volumes. He went back to London when he retired, and served as Secretary of the British Ornithologists' Union from 1898-1901. Dr. W.T. Blanford was a member of the Indian Geological Society, and during his many journeys, kept a keen eye on the fauna of British India. He also contributed to the various journals.
1922-30	'Fauna of British India including Ceylon and Burma: Birds' (2nd edition in 7 volumes)	E.C. Stuart Baker Baker was a British ornithologist and a police officer in Indian Police Service. He spent most of his career in India in the Assam Police, rising to the rank of Inspector-General. He was an enthusiastic big game hunter. He lost his left arm to a panther in Silchar, Assam. Later, for his services during the First World War, he was appointed Officer of the Order of the

		British Empire (OBE). He was a contributor to *Ibis,* and the *Journal of Bombay Natural History Society.* He edited the work by Oates and Blanford, using a trinomial system of nomenclature, so that geographical races and subspecies were delineated from species.
1918-22	'A Monograph of the Pheasants' (in 4 volumes)	Dr. Charles William Beebe Dr. Beebe was an American naturalist, explorer, and author. In 1899, he was appointed Curator of Ornithology for the New York Zoological Society, and later in 1919, the New York Zoological Society's Director of the Department of Tropical Research. In addition to the United States and Canada, he eventually undertook expeditions in Trinidad, Venezuela, Brazil, and Asia. Beebe wrote many popular books of his expeditions (*Pheasant Jungles, Edge of the Jungle, Jungle Days*), some of which became best-sellers in the 1920s and 1930s. He was also a regular contributor to the *National Geographic Magazine.* He wrote *A Monograph of the Pheasants* (1918–1922) in 4 volumes, which remains the classic reference on the subject. It is the outcome of Beebe's 17-month journey through the various countries of Asia where pheasants are found.

Fig. 2.11 Cover of the book *A Monographs of the Pheasants*, by William Beebe, created in 1918-22.

Naturalists who have studied and written about pheasants of China and Indo-China include British Consul Robert Swinhoe 1836-1877 (his brother Col. Charles Swinhoe was one of the founders of the Bombay Natural History Society), and lawyer and zoologist P.L. Sclater (1829-1913), who was the founder of the journal *Ibis*. The Swinhoe's pheasant and Sclater's monal are named after them. C. J. Temminck (1778–1858) was a Dutch aristocrat and zoologist. Temminck was the first director of the National Natural History Museum at Leiden. His *Manuel d'ornithologie, ou Tableau systematique des oiseaux qui se trouvent en Europe* (1815) was the standard work on European birds for many years. He inherited a large collection of bird specimens from his father, who was treasurer of the Dutch East India Company. In 1831, he was elected a foreign member of the Royal Swedish Academy of Sciences. Temminck was also the author of *Histoire naturelle générale des Pigeons et des Gallinacées* (1813–1817), and about 10 other books. Many species of fish, reptiles, birds, and mammals are named after him, including the Temminck's Tragopan. Hon. Walter Rothschild (1868-1937) was a

British banker, politician, and zoologist. His collection of animals is housed at the Natural History Museum in England. Many species of animals are named after him, including the Rothschild peacock pheasant (Wikipedia).

Fig. 2.12 A painting of a pair of blood pheasants by Joseph Wolf from *A Monograph of the Phasianidae* by D G Elliot, 1872.

Daniel Giraud Elliot (1835 - 1915) was an American zoologist. Elliot was one of the founders of the American Museum of Natural History (in New York) and the American Ornithologists' Union. He was appointed curator of zoology at the Field Museum in Chicago in 1894. In 1872, D.G. Elliot prepared the first monograph of the pheasants, a work which contained some exquisite hand-coloured reproductions of watercolours, painted based on his private specimen collection (collected on his trips to Asia), by Joseph Wolf. [Elliot also published a series of beautiful color-plate books on birds and animals, for a small select list of subscribers. Elliot wrote the text himself, and commissioned artists Joseph Wolf and Joseph Smit to provide the illustrations. The books included - *A Monograph of the Phasianidae* (1870-72), *A Monograph of the Paradiseidae* (1873), *A Monograph of the Felidae* (1878), and *Review of the Primates* (1913)]. Although not all the pheasant species had been described when this work was published, the book is a marvelous celebration of the beauty by which this group of birds has entranced man. The National Academy of Sciences (Washington DC, USA) constituted an award known as the Daniel Giraud Elliot medal in 1917 "for meritorious work in zoology or paleontology published in a 3- to 5-year period". American naturalist Dr. William Beebe was awarded this medal in 1918; and American ornithologist Dr. James Paul Chapin (leader of the Lang-Chapin expedition which made a biological survey of the Belgian Congo between 1909 and 1915) was conferred with this medal in 1932 (Wikipedia). Interestingly, the African pheasant, Congo peacock *Afropavo congensis*, was unknown to science at the time Dr. Beebe wrote his *Monograph*, and thus it does not contain a description of this species. Congo peacock was described years later in 1936, after it was accidentally 'discovered' in a museum in Belgium. In 1936, Dr. James Chapin described it as a species, based on two stuffed specimens at the Royal Museum for Central Africa in Tervuren, Belgium (Congo was King Leopold II of the Belgians' territory from 1885-1908, and under Belgian colonial rule until 1960). From an earlier expedition, Chapin had collected native Congolese headdresses containing long feathers of an unidentified bird. Since the Congo peafowl is very sparsely distributed throughout its native range in the lowland rainforests of Congo, Dr. Chapin's survey did not detect it. When he went to the museum in Belgium in 1934, he found 2 stuffed specimens of the species, incorrectly labeled as 'Indian Peacocks'! These specimens matched the feathers found in the headdresses collected from Congo. So, in 1937, Chapin went to Congo and collected a few specimens of the Congo peacock (known locally as 'mbulu'), though he officially described the species a year earlier (see 'Further Reading' at the back).

Another type of literature on pheasants published in 19th and early 20th century was the books on aviculture and hunting of the introduced species of pheasants in North America and England. English naturalist and a friend of Charles Darwin, Mr. W. B. Tegetmeier, wrote *Pheasants: Their Natural History and Practical Management* (1881). Mr. Frank Finn,

English ornithologist and Deputy Superintendent of the Indian Museum, Kolkata, wrote more than 10 books on birds, such as *Fancy Pheasants and their Allies* (1901) and *Indian Sporting Birds* (1915, coauthors A. O. Hume and C. H. T. Marshall). Scottish ornithologist and Curator in British Museum (now Natural History Museum), Mr. W. R. Ogilvie-Grant wrote *A Hand-Book to the Game Birds* in 1896.

Fig. 2.13 A book by Dr. P. A. Johnsgard, published in 1999 by the Smithsonian Institution Press, Washington, D.C.

The American ornithologist Jean T. Delacour, in 1951 published his authoritative work, *The Pheasants of the World*, covering in detail, for the first time, all known species of pheasants. A revised and updated second edition was published in 1977. Mr. P. Wayre, who founded the Pheasant Trust in UK in 1959, published *A Guide to the Pheasants of the World* in 1969 which gave a good general account of pheasants and their management. Two ornithologists, Indian Dr. Salim Ali and American Dr. S. D. Ripley (Smithsonian Institution, USA), in 1969, produced the comprehensive series called *Handbook of the Birds of India and*

Pakistan, dealing with all the birds found in this geographical area. The 2nd volume contains information about the order Galliformes, and almost all the species are illustrated in color. It has recently been revised (in 1980). Keith C. R. Howman (former President of WPA UK) and American ornithologist Dr. Paul A. Johnsgard (University of Nebraska, USA) have also published interesting and informative works.

Some of the Indian wildlife scientists who have done pioneering research on pheasants in the past few decades are –

Dr. S. Sathyakumar, Dr. Rahul Kaul, Dr. Anwarrudin Choudhury, Dr. Dipankar Ghose, Dr. Rajiv S. Kalsi, Dr. K. Ramesh, Dr. M.C. Sathyanarayana, Dr. K. Sivakumar, Dr. Sarla Khaling, Dr. Ashfaque Ahmad, Dr. Shahla Yasmeen, and Dr. M. Shah Hussain. They are a part of the institutes and organizations, such as Wildlife Institute of India, Aligarh Muslim University (Department of Wildlife Sciences), World Pheasant Association, and Wildlife Trust of India – that form the network of galliform research and conservation along with university and government departments in India.

WPA scientists from the UK who have been associated with research projects in India are - Dr. Peter J. Garson and Dr. Philip McGowan. Pheasant breeders and experts, and WPA members Mr. John Corder and Mr. Francy Herman have also made significant contributions to *ex situ* conservation of pheasants.

Chapter 3

The 50 Species of Pheasants

I chose Sandukphu as a halting place from which to make excursions west into Nepal, north to Phallut and on to the gorge of the Changthap, where Dhanga La rises to a cold and barren 15000 feet, and the eternal snows of the outer spurs of Kinchinjunga are only ten miles away. In this region I found and studied the blood pheasants, the impeyans and the tragopans.

- From *Pheasant Jungles* by William Beebe

Fig. 3 Painting of a pair of blood pheasants *Ithaginis cruentus*, by H C Richter in *A Century of Birds from the Himalayas* (1830-32) by John Gould.

Chapter 3

The 50 Species of Pheasants

1. **Genus *Ithaginis* – The Blood Pheasant**

This genus has one species - *Ithaginis cruentus*.

The blood pheasant is a montane species. The sexes are moderately dimorphic (distinguishable at sight/ in the field as male or female). In males, the plumage is soft, grayish above, apple-green streaked with yellow below. The males are colored crimson on the chin and the rump. A short crest (clump of feathers on the head), and a red orbital patch (featherless skin patch around the eyes) is present in both sexes. Females are colored a rich rufous-brown all over.

Table 3.1 Species statistics at a glance.

Species	Altitude Range	Countries of occurrence of the Species	Population in the Wild	Population Trend
Ithaginis cruentus (Blood pheasant)	1830 – 4575 m	India, Nepal, Bhutan China, Burma	unknown	unset

1. **Species: *Ithaginis cruentus* – Blood pheasant**

Local names: *Semo* (Tibetan), *Chilime* (Nepalese).

Description: In males, the plumage is soft, greyish above, apple-green streaked with yellow below. The males are colored crimson on the chin, upper breast and the rump. A short crest (clump of feathers on the head) and a red orbital patch (featherless skin patch around the eyes) is present in both sexes. Females are colored a rich rufous-brown all over. Short, stout, curved beak.

Call: Alarm note 'ship, ship' and a repeated high-pitched 'chic', and a long, high-pitched squeal resembling that of a kite. Trilling 'sree cheeu, cheeu, cheeu'.

Habitat: Open coniferous forests of firs and oaks, higher shrub-like zone of rhododendrons, bamboo jungle close to snowline, and alpine meadowland. Forage in the open areas during the day, and roost on rhododendrons or firs at night.

Food: Berries, insects, seeds, shoots of ferns and pines, mosses, lichens, and bamboo leaves.

IUCN Red List Category: Least Concern.

2.   Genus *Tragopan*– The Horned Pheasants

This genus has five species - *Tragopan melanocephalus, T. satyra, T. temminckii, T. blythii* and *T. caboti*.

The tragopans are 'horned' montane pheasants. The sexes are highly dimorphic (distinguishable at sight as male or female).

The plumage in males is crimson red with extensive white/ buffy spotting. The males have 2 erectile and brightly coloured fleshy horns on the head and, a brightly coloured expandable lappet at the neck that is only visible during courtship. Orbital patch (featherless skin patch around the eyes) is present in both sexes; each species has a different colour. Females are colored a grey-brown all over, the 5 species differing slightly in shades and patterns. Short, stout beak for digging out bulbs and roots.

Table 3.2 Species statistics at a glance.

Species	Altitude Range	Countries of occurrence of the Species	Population in the Wild	Population Trend
Tragopan melanocephalus (Western tragopan)	1750-3600 m	Pakistan, India	5,000	decreasing
T. satyra (Satyr tragopan)	1800-4250 m	India, Nepal, Bhutan, Tibet	10,000-19,999	decreasing
T. temminckii (Temminck's tragopan)		Tibet, Burma, China	100,000-499,999	unset
T. blythii (Blyth's tragopan)	1800-3300 m	India, Tibet, Bhutan, Burma, China	2,500-9,999	decreasing
T. caboti (Cabot's tragopan)	600-1800m	China	2,500-9,999	decreasing

2.   Species: *Tragopan melanocephalus* – Western tragopan

Local names: *Jujurana* (king of birds- Kullu), *Pyara* (kinnaur), *Jyazi* (Bushahr), *Fulgar* (Chamba and Kangra).

Description: In males, the plumage is crimson and black, spotted white. Red orbital patch (featherless skin patch around the eyes), deep blue throat. The male's lappet is bluish purple centrally, with 4-5 reddish, irregular to leaf-like markings along each side. Females have a grey-brown plumage, with black and white streaks and spots on upperbody. Short, stout beak.

Call: A rather goat-like '*waa-waa-waa*' , at day-break and dusk. Territorial calling in early April when the males scatter and begin to pair.

Habitat: Coniferous or mixed mountain forests with dense undergrowth, or rhododendron scrub. Roost on low evergreen trees at night. Very shy and elusive. During conditions of heavy snowfall, they may occasionally emerge to feed in the open in groups of 2-3.

Food: Fresh leaves, berries, seeds, flowers, roots, bamboo shoots and insects.

IUCN Red List Category: Vulnerable

✠ Major causes: Habitat degradation and fragmentation through subsistence farming, browsing of understorey shrubs by livestock, tree-lopping for animal fodder and fuelwood-collection are the main threats. Disturbance by graziers and particularly collectors of edible fungi and medicinal plants may seriously interfere with nesting. Hunting and trapping for its meat (especially in winter) and its decorative plumage pose additional threats throughout Pakistan, and Chamba, Himachal Pradesh (India).

Conservation measures underway- It is afforded legal protection in both India and Pakistan. It occurs in three national parks: Machiara (Pakistan), and Kishtwar, and the Great Himalayan (India), and also 10 Wildlife Sanctuaries. Discovery of the large Palas population triggered a major conservation initiative in the region for which this bird is the flagship species. A galliform monitoring and conservation project continues within the valley. Surveys have been conducted recently across most of its presumed range in Pakistan, and in Himachal Pradesh where in 2005, 3000 Forest Guards and Officers were involved in a coordinated week long state-wide survey, now conducted annually. It is currently the subject of a large conservation breeding programme in Himachal Pradesh, with the potential for future releases of parent-reared offspring to augment/restock local wild populations.

3. Species: *Tragopan satyra* – Satyr tragopan

Description: Male's plumage is nearly blood red , with white markings having black edging on back and below. Bluish face and throat. Orbital skin (featherless skin patch around the eyes) and lappet are Prussian blue with patches of brilliant red (lappet 3.5 inches long). Females generally rufous brown with sandy or blackish mottles above and white spots below. Short, stout beak.

Call: Male courtship call is a loud '*wak*' or '*kya*' during April-May. Also, a deep full-throated '*wah, waah! oo-ah ! oo-aaaaa!*

Habitat: Moist and dry temperate coniferous mountain forests with dense undergrowth, rhododendron forest and, gorges of streamlets, springs on steep hillsides. Roost 3m up on trees at night. Very shy and elusive. Mostly solitary, forage during early morning and late afternoon on the open edges of the forest.

Food: Fresh leaves, berries, seeds, rhododendron flowers, bulbs, ferns, bamboo shoots and insects.

IUCN Red List Category: Near Threatened

✳ Major causes: Major threats include excessive hunting; it is occasionally snared by local people for food, as well as habitat clearance and degradation due to timber harvesting, fuel-wood and fodder collection and livestock grazing.

4. Species: *Tragopan temminckii* – Temminck's tragopan

Local names: *Kiao-ky* (Chinese)

Description: Male's plumage is nearly blood red, with grayish spotting on the flanks and underparts lacking the black edging of the satyr. Light blue orbital skin (featherless skin patch around the eyes) and, lappet is blue, mottled with lighter blue spots centrally, bright red markings along each side. Females very similar to Satyr. Short, stout beak.

Call: Male courtship call is a series of '*waaa*' notes. Challenge call *'ona ona'*.

Habitat: Found in dense montane mixed deciduous and coniferous forests on steep slopes, with thick undergrowth and moss. Highly arboreal. Forage singly, in pairs or small family groups.

Food: Fresh leaves, berries, fruits, flower buds and insects.

IUCN Red List Category: Least concern

5. Species: *Tragopan blythii*- Blyth's tragopan

Description: Male's head, neck and chest are orange red. Almost uniformly gray below. Females distinctly grayish on the undersides, dark olive brown above. Golden yellow orbital skin (featherless skin patch around the eyes). Male's lappet is small (2-2.9 inches), rounded, mostly yellow, bordered with pale blue. Erectile horns are pale blue. Short, stout beak.

Call: Male courtship call is *'wak'* or *'wa ak-ak'.* Challenge call *'mao mao'.*

Habitat: Found in moist temperate and evergreen coniferous forest with thick undergrowth. Trees draped in epiphytes (mosses, orchids), near seasonally turbulent mountain streams. Forage in pairs or small parties of 4-5 birds.

Food: Fresh leaves, berries, fruits, flower buds, ferns, seeds, and insects.

IUCN Red List Category: Vulnerable

⊠ Major causes: In north-east India, deforestation is a significant threat, primarily as a result of shifting cultivation. Together with fuelwood-collection and commercial timber extraction, this is rapidly fragmenting suitable habitat, even within protected areas, where enforcement of regulations is often absent or impossible. Hunting for food is the other major threat, particularly in Nagaland and Arunachal Pradesh, where large-scale snaring of pheasants and partridges by local people is an increasing problem. Little data on the exploitation of this species is available from Myanmar, making it difficult to assess the severity of the threat there. Even in Bhutan, high levels of grazing and slash-and-burn agriculture are potentially significant problems.

Conservation measures underway- The species is legally protected in all countries. It occurs in several protected areas, including: two small wildlife sanctuaries and a community reserve in Nagaland; the Blue Mountain National Park in Mizoram; Mouling National Park, Sessa Orchid Sanctuary, Eaglenest Mehao and Dibang Wildlife Sanctuaries in Arunachal Pradesh; Thrumsing La National Park, Bhutan; Gaoligongshan National Park in China; and Natma Taung National Park, Myanmar. Surveys for the species have been conducted in many areas in north-east India. An international studbook exists documenting the captive population held at locations in North America and Europe; however, recent analysis found the captive population is declining, ageing and highly inbred and requires new founders if it is not to be lost as a conservation resource for the species.

6. Species: *Tragopan caboti* – Cabot's tragopan

Description: Male has plain buff underparts without any ocelli, the upper parts have only indistinct buffy spots that are flanked by black and russet. Orbital skin is orange yellow. Throat pale blue. Erectile horns 2 inches long, pale blue. Lappet is large (5.9 inches), orange centrally, purple spots on lower half which is cobalt blue with scarlet patches along margins. Females are rofous brown above, with whitish triangular marking. Grayish-brown below, with whitish spotting. Reddish to orange orbital skin. Short, stout beak.

Call: Loud repeated 15-16 '*chi*' notes during display sequence.

Habitat: Found in moist montane mixed forests having dense undergrowth dripping with moisture. Forage in pairs or family units.

Food: Fresh leaves, acorns, fruits, seeds and mollusks.

IUCN Red List Category: Vulnerable

Major causes: The main threat is habitat loss and modification, particularly where this leads to habitat fragmentation. Its recorded limited ability to disperse across gaps in forest cover greater than 500 m means the species is highly susceptible to fragmentation. Most natural forest has been cleared or modified as a result of the demands for agricultural land and timber. The progressive replacement of natural evergreen broadleaf forests with conifer plantations and bamboo is now a major problem for this species, in part because the species nests on natural platforms such as forks between branches, which are lacking in conifer trees: provision of artificial nesting platforms may allow the species to persist in conifer habitats. Illegal hunting for food still occurs in some places, especially outside protected areas. Since 1983, 48% of known nests have been destroyed by natural predators, in particular Eurasian Jay *Garrulus glandarius*, Yellow-throated Marten *Charronia flavigula* and Leopard Cat *Felis bengalensis*.

3.   Genus *Pucrasia* – The Koklass Pheasant

This genus has one species, *Pucrasia macrolopha*.

The koklass pheasant is a montane species (live in mountainous habitats). The sexes are moderately dimorphic (distinguishable at sight as male or female). The males have a well developed occipital crest and erectile neck feather tufts. The females have a shorter crest and lack the tufts. Head is entirely feathered. Tail is highly graduated.

Table 3.3 Species statistics at a glance.

Species	Altitude Range	Countries of occurrence of the Species	Population in the Wild	Population Trend
Pucrasia macrolopha (Koklass pheasant)	2000-4000m	Afghanistan, India, Nepal, Tibet, China, Mongolia	unknown	unset

7. Species: *Pucrasia macrolopha* – Koklass pheasant

Local names: *Plash, khwaksha* or *koklas* in Shimla district; *khwaksha* in Kinnaur district.

Description: The male has a greenish black head, chestnut colored breast, white patches on the sides of the neck. Somewhat elongated blackish to brownish tail with paler tip. Tapering head crests, in males, raised into earlike display structures. The females also have whitish patch along the sides of the neck.

Call: Loud 'pok-*pok-pok*' during morning and evening hours.

Habitat: Found in oak and coniferous forests and scrub zone. Forage singly or in pairs. Roost 6-9 m high on pine trees.

Food: Needles of pine, acorns, berries, buds, insects, moss, and grass.

IUCN Red List Category: Least Concern

4.   Genus *Lophophorus* – The Monals

This genus has three species - *Lophophorus impeyanus, L. sclateri* and *L. ihuysi.*

The monals are large montane pheasants.   The sexes are highly dimorphic (distinguishable at sight as male or female). Iridescent bluish purple plumage is extensive in the males; under-parts velvety black. The males have a bright blue orbital patch and crests of different shapes. The beak is highly curved, the upper mandible overlapping the lower one.

Table 3.4 Species statistics at a glance.

Species	Altitude Range	Countries of occurrence of the Species	Population in the Wild	Population Trend
Lophophorus impeyanus (Himalayan monal)	2100-4500 m	Afghanistan, Pakistan, India, Bhutan, Tibet, Burma	unknown	unset
L. sclater (Sclater's monal)	2000-4200 m	Tibet, Burma	2,500-9,999	decreasing
L. ihuysi (Chinese monal)	2800-4500 m	Tibet, China	10,000 - 20,000	decreasing

8. Species: *Lophophorus impeyanus* – Himalayan monal

Local names: *Bnal* (male), *Bodh* (female) or *Kardi* (female) – in Shimla and Kullu districts; *Dang* in Kinnaur district of Himachal Pradesh.

Description: The male is a mix of glossy, metallic purple, blue and bronze-green above; prominent crest of iridescent green feathers; white lower back and rump; chestnut wings distinctive in flight; short rufous tail; velvety black underparts. The female is mostly a chestnut brown, with buffy body streaks; light blue skin area around the eye. Also has an erectile occipital crest.

Call: Loud ringing whistle; a '*chuck chuck*' alarm call.

Habitat: Occur in open coniferous and mixed forests, rocky slopes, gorges, grassy areas interspersed with woods. Forage in small groups.

Food: Tubers, roots, seeds, berries, and insects.

IUCN Red List Category: Least Concern

9. Species: *Lophophorus sclateri* – Sclater's monal

Description: The males of this species resemble Himalayan monal, but have much more white on the lower back, rump and tail; chestnut tail is white-tipped. Short crest. Females have a paler lower back, rump and tail than the Himalayan monal. Have a buffy tail band.

Call: A shrill harsh whistle.

Habitat: Occur in dense montane forests having a rhododendron understory and grassy openings. Roost and feed in small groups. Forage in small forest openings during morning and evening.

Food: Tubers, roots, seeds, buds.

IUCN Red List Category: Vulnerable

🦚 Major causes: Hunting for food is the single main threat across the species's range. In addition, hunting for feathers (to make ornaments and fans) is a localised problem in India. In the Arbor Hills and Mishmi Hills, India, hunting intensity has significantly reduced population densities. Habitat degradation as a result of logging is a more localised threat. The habitats of the newly discovered taxon in Arunachal Pradesh appear to be little threatened, owing to their inaccessibility.

Conservation measures underway - It occurs in the Gaoligong Shan National Nature Reserve in Yunnan, China, although it appears to be rare here and numbers are probably small. The creation of Dehang-Debang Bioshpere Reserve in Arunachal Pradesh, which encompasses the Dibang Valley Wildlife Sanctuary, Mouling National Park and unclassified state forests, offers further protection. The remote nature of its habitats and range, to an extent, lessen the level of threat it faces from hunting. A new subspecies was recently described and its conservation status has been reviewed.

10. Species: *Lophophorus ihuysi* – Chinese monal

Local names: *Koa-loong* (Tibetan Chinese), *be-mu-chi* (Chinese)

Description: The males do not have a chestnut colouration on the tail. The females have a larger, more contrastingly patterned white back. Resemble other monal females.

Call: Loud '*koa-loong*' in early morning. A repeated loud, gabbling alarm whistle.

Habitat: Occur in coniferous forests having a rhododendron understory; alpine meadows ; above tree line. Occur in small parties, foraging in open habitats; at night perch on rhododendrons or pine branches.

Food: Bulbs, roots, flowers, tender leaves, sprouts.

IUCN Red List Category: Vulnerable

Major causes: Its subalpine and alpine meadow habitats have been degraded in some areas by an increase in the grazing of wild yaks. The large-scale collection of *Fritillaria* spp. (a known food of this species) and other herbs for Chinese medicine causes local disturbance, and nests are sometimes destroyed by these activities. Illegal hunting is also considered to be a localised threat, and appeared to be the cause of a substantial decline at Baoxing where this species was surveyed in 1983-1986, and again in 1988. Reports suggest that hunting pressure on the species has increased in recent years. The forests in west Sichuan have been rapidly exploited in recent decades, and that has directly affected its subalpine habitats, and logging roads have improved access to alpine habitats for local people.

5.   Genus *Gallus* – The Junglefowl

This genus has four species - *Gallus varius, G. gallus , G. sonnerati* and *G. lafayettei*.

The junglefowl are tropical pheasants. The sexes are highly dimorphic (distinguishable at sight as male or female). The males have iridescent plumage, a fleshy comb on the head, and 1 or 2 lappets below the bill. In females the plumage is dull, combs and lappets are reduced in size.

Table 3.5 Species statistics at a glance.

Species	Altitude Range	Countries of occurrence of the Species	Population in the Wild	Population Trend
Gallus varius (Green junglefowl)	0-2000m	Java, Bali	unknown	unset
G. gallus (Red junglefowl)	0-2000m	Pakistan, India, Burma, China, Java, Sumatra, Bali, Philippines, Cambodia, Vietnam, Laos, Thailand	unknown	unset

G. sonnerati (Sonnerat's/Grey junglefowl)	0-1500m	India	unknown	unset
G. lafayettei (Ceylon junglefowl)	0-2000m	Sri Lanka	unknown	unset

11. Species: *Gallus varius* – Green junglefowl

Description: The male's plumage is greenish to blackish; iridescent rounded green neck feathers; yellow edged rump; a rounded green and purplish comb. Females dull; dark tail with iridescent green sheen; spotted or scaly back.

Call: Male crowing call shrill, crisp *'chaw-aw-awk'* or *'chak'* or *'kowak'*; alarm call repeated *'chop'*. Female cackling *'wuk'*; alarm call *'kowak kowak'* or *'kok'*.

Habitat: Tropical seashores, coastal valleys, inland forests, bamboo thickets.

Food: Insects, seeds, berries.

IUCN Red List Category: Least Concern

12. Species: *Gallus gallus* – Red junglefowl

Description: The male has golden brown to reddish neck feathers; a serrated scarlet comb; blackish green tail and under-parts. Female dull brown with yellowish and blackish pattern.

Call: Male crowing call at morning and evening *'cock-a-doodle-doo'*.

Habitat: Tropical to subtropical habitats, flat or rolling terrain; edge habitats; Mixed evergreen and deciduous forests; dense undergrowth or bushy habitat of bamboo, lantana, *Ziziphus*. Roost socially in trees; forage in the morning and evening hours.

Food: Insects, seeds, fruits, leaves, snails.

IUCN Red List Category: Least Concern

13. Species: *Gallus sonnerati* – Grey junglefowl

Description: The male is grayish in appearance; neck, rump and wing coverts have white and yellow spots. Female plumage dark brown with black and white spotted, mottled under-parts.

Call: Male crowing call at morning and evening '*kuck-kaya-kaya-kuck*', '*kyukun, kyukun*'.

Habitat: Dry deciduous to moist evergreen forests, mixed bamboo jungles. Found in pairs or small family groups.

Food: Shoots, tubers, berries, figs, seeds of bamboo, insects, and small reptiles.

IUCN Red List Category: Least Concern

14. Species: *Gallus lafayettei* – Ceylon junglefowl

Local names: *Wali-kukula* (Native Ceylonese).

Description: The male has brown streaking on yellow neck; iridescent bluish black wings; reddish brown underparts; comb is less serrated than G. gallus, inner part of comb is yellowish. Females have strongly barred wing and tail pattern.

Call: Male crowing call '*chick, chow-choyik*' (or ' *tsek ....George Joyce*').

Habitat: Forests, bamboo thickets. Forage in the morning and evening, females and their brood flock.

Food: Seeds, berries, insects, mollusks.

IUCN Red List Category: Least Concern

6. Genus *Crossoptilon* – The Eared Pheasants

This genus has three species - *Crossoptilon crossoptilon, C. auritum, C. mantchuricum*. A fourth species, *C. harmani* (Tibetan eared pheasant) has been proposed.

The eared pheasants are montane pheasants. Sexual dimorphism is lacking (cannot be distinguished at sight as male or female).

Both sexes have elongated ear-feathers that form 'horns' or 'ears' on the sides of the head. The rest of the plumage is somewhat hair-like. Iridescent plumage is virtually lacking in both sexes. The ear-feathers are white. Red orbital patch present.

Table 3.6 Species statistics at a glance.

Species	Altitude Range	Countries of occurrence of the Species	Population in the Wild	Population Trend
Crossoptilon crossoptilon (White eared pheasant)	3000-4300 m	Tibet, China	10,000 - 50,000	decreasing
C. auritum (Blue eared pheasant)		China, Mongolia	unknown	unset
C. mantchuricum (Brown eared pheasant)	1100-2600 m	Mongolia	5,000 - 17,000	decreasing

15. Species: *Crossoptilon crossoptilon* – White eared pheasant

Local names: *Sharkar* (Tibetan).

Description: Males and females have white body plumage, darker tail plumage.

Call: Alarm call '*wrack*', or '*cuco, cuco*'. Breeding call of male '*kak*' and '*trip-crrra-ah*'. Female responds '*krkrkr-krkrkr*'.

Habitat: Subalpine forests near snowline (upper limits of coniferous mixed forest); rhododendron and juniper scrub; grassy slopes.

Food: Bulbs, berries.

IUCN Red List Category: Near Threatened

Major causes: It is threatened by hunting for food and deforestation, but the high-altitude forests that it inhabits are not being lost at a very rapid rate. Completion of a rail link to Tibet is imminent and will lead to increased tourism in the area. It is feared that knock-on effects of this will increase the rate of habitat loss.

16. Species: *Crossoptilon auritum* – Blue eared pheasant

Local names: *Ho-ki* (Chinese)

Description: Males and females mostly smoke gray; paler areas on the rump; white on the base of outer tail feathers. Red orbital patch present.

Call: Breeding call of male 'ka-ka...la', 'krip-krraah-krraah', in early mornings.

Habitat: Coniferous and mixed forests; bushy sites on alpine meadows above the forest. Quite social; form large flocks during the non-breeding season. Forage during early morning or late afternoon. Return to tree cover in the evenings to roost 2-4 meters above ground.

Food: Bulbs, leaves, roots, insects.

IUCN Red List Category: Least Concern

17. Species: *Crossoptilon mantchuricum* – Brown eared pheasant

Local names: *Ho-ki* (Chinese)

Description: Males and females mostly brown; white rump and anterior tail areas. Long white ear-tufts. Red orbital patch present.

Call: Breeding call of male 'Trip-c-r-r-r-r-ah'.

Habitat: Subalpine forests of birches, oaks, pines.

Food: Bulbs, tubers, seeds, leaves, roots, insects, earthworms.

IUCN Red List Category: Vulnerable

�֍ Major causes: Its range has been fragmented by habitat loss and isolated populations are at risk from further forest loss and other pressures. Outside nature reserves, the threats include deforestation for agriculture and urban development, and habitat degradation due to logging and livestock-grazing. Local people collecting fungi may be the cause of high nest failure rates at Pangquangou National Nature Reserve.

7.   Genus *Lophura* – The Gallopheasants

This genus has ten species - *Lophura inornata, L. imperialis, L. edwardsi, L. leucomelana, L. nycthemera, L. swinhoei, L. diardi, L. erythropthalma, L. ignita, L. bulweri.* 2 more species have been proposed – *L. hatinhensis* (Vietnamese fireback) and *L. hoogerwerfi* (Sumatran/Aceh pheasant).

The gallopheasants are tropical to montane pheasants. The sexes are dimorphic (distinguishable at sight as male or female). The males typically have extensive purplish to greenish iridescence on upper body; blackish underparts; varying amounts of white on the tail, back and head. Erectile red or blue velvety wattles around the eyes of males, and similar orbital areas in females. Females are brown to blackish with varying amounts of spotting or barring.

Table 3.7 Species statistics at a glance.

Species	Altitude Range	Countries of occurrence of the Species	Population in the Wild	Population Trend
Lophura inornata (Salvadori's pheasant)	650-2200 m	Sumatra	2,500-9,999	decreasing
L. imperialis (Imperial pheasant)		Vietnam, Laos		
L. edwardsi (Edward's pheasant)	Up to 300 m	Endemic to Vietnam	250-999	decreasing
L. leucomelanos (Kalij pheasant)		India, Nepal, Bhutan, Bangladesh, Burma, Thailand, (Introduced to Hawaiian islands in 1962)	unknown	unset
L. nycthemera (Silver pheasant)		China, Burma, Cambodia, Thailand, Vietnam, Laos	unknown	unset
L. swinhoei (Swinhoe's pheasant)	200-2300 m	Taiwan	5,000 - 10,000	decreasing

L. diardi (Siamese/Diard's fireback)	500-1150 m	Vietnam, Laos	5,000 - 10,000	stable
L. erythropthalma (Crestless fireback)		Borneo, Sumatra, Malaysia	10,000-19,999	decreasing
L. ignita (Crested fireback)		Borneo, Sumatra, Malaysia	unknown	decreasing
L. bulweri (Wattled pheasant)	200-1500 m	Endemic to Borneo	2,500-9,999	decreasing

18. Species: *Lophura inornata* - Salvadori's pheasant

Description: Males almost entirely bluish black; crest absent. Females chestnut brown, buffy mottling, blackish brown tail. Red orbital patch in both sexes.

Habitat: Montane forests of Sumatra.

IUCN Red List Category: Vulnerable

❉ Major causes: Hunting pressure is thought to have caused declines in parts of the species's range, but it still occurs close to heavily settled areas, and thus appears to be resilient to a degree of trapping. In 1999 a congener *L. hoogerwerfi* was recorded in bird markets in Medan, northern Sumatra for the first time; the extent to which *Lophura* pheasants feature in national or international trade is not known. Much of the forest within the lower part of the species's altitudinal range around Kerinci has already been cleared for shifting cultivation, and is vulnerable to further illegal agricultural encroachment and increasingly frequent drought fires. The range is becoming increasingly fragmented, a trend which is likely to continue.

19. Species: *Lophura imperialis* - Imperial pheasant

Description: Males entirely dark blackish; short and dark crest present. Females similar to Edward's, slight differences in coloration. Crest present. Red orbital patch in both sexes. Some scientists regard it as a hybrid of Edward's pheasant and Silver pheasant.

Habitat: Secondary lowland forests.

Food: Insects, seeds, plants.

20. Species: *Lophura edwardsi* - Edward's pheasant

Description: Male has white crown and crest. Facial skin is red. Female similar to Imperial.

Call: '*Chuck*' crowing call in spring.

Habitat: Level to gently sloping dense coastal-plain moist forests, closed canopies, interspersed bamboo patches.

Food: Seeds, plants.

IUCN Red List Category: Endangered

✖ Major causes: Its historical range is now almost completely denuded of primary forest through a combination of herbicide spraying during the Vietnam War, logging and clearance for agriculture. The last forest areas known to support the species are subject to continuing degradation by wood-cutters. It has been suggested that habitat disturbance might increase following the completion of the Ho Chi Minh National Highway which bisects the three provinces in which the species is known to occur. Hunting pressure from various forest-product collectors poses an additional threat and the species may be affected by indiscriminate snaring.

21. Species: *Lophura leucomelanos* - Kalij pheasant

Local names: *Kalesa, Kansha* or *Kolsa* in Shimla, Solan. *Panti* or *Kulsus* in Kinnaur.

Description: Male has a red orbital patch, occipital crest, black tail. Female is also crested, has brown plumage. While the subspecies *L. l. hamiltoni* is found in western Himalayas, *L. l. lathami* is found in northeast India, eastern Bhutan and Burma. There are 7 more subspecies of this pheasant.

Call: Alarm call repeated '*whoop keet keet*'. Low conversational '*kurr-kurr-kurrchi-kurr*'.

Habitat: Dense evergreen and deciduous forests near streams. Tropical and montane forest habitats. Forage in pairs or family groups. Roost on trees at night, 6-12 meters above ground.

Food: Bamboo seeds, small snakes, lizards, termites, figs, roots, tubers, berries.

IUCN Red List Category: Least Concern

22. Species: *Lophura nycthemera* - Silver pheasant

Description: The male has a thick black decumbent crest, white neck, black underparts, a very elongated tail that is white centrally and barred with black and white laterally. Red orbital patch. Females have a distinct crest on head; tail patterned with dark brown and black lines on the sides.

Habitat: Damp mountain forests; broad areas of grassland that are bordered by forests; hilly areas; foothills bamboo woods and evergreen forests. Roost in trees, foraging at dusk and dawn in groups of 3-7.

Food: Insects, flower petals, leaves, berries, seeds, small reptiles, worms, roots, tubers.

IUCN Red List Category: Least Concern

23. Species: *Lophura swinhoei* - Swinhoe's fireback pheasant

Local names: *Wa-koe* (Chinese)

Description: Males have a white crest; iridescent purplish black plumage; white central tail feathers; white mantle. Red facial skin present. Females lack crest.

Call: Alarm call '*deek*'. Display call '*tse tse*', '*check check*'.

Habitat: Montane evergreen tropical forests with scattered light scrubs and ferns as understorey. Coniferous and mixed forest. Gentle slopes. Forage in open areas of the forest floor during morning and late afternoon hours. Roost on trees.

Food: Insects, flower buds, leaves, berries, earthworms.

IUCN Red List Category: Near Threatened

Major causes: Heavy hunting pressure was a problem for it in the past, but is no longer a serious threat. Deforestation is still a threat outside protected areas; there is the risk that subpopulations will become isolated in the future.

24. Species: *Lophura diardi* - Siamese fireback pheasant

Local names: *Kaipha* (Siamese)

Description: The males have a comma-shaped crest, scarlet facial skin; a small patch of maroon on the back. The female has rufous colored plumage; buffy white barring on upper wings and central tail feathers; lacks crest.

Call: A whistling call, loud repeated '*pee-yu*'.

Habitat: Lowland evergreen forests.

Food: Insects, fruits , berries, crabs. Forage singly or in groups.

IUCN Red List Category: Near Threatened

Major causes: This species is threatened by continuing extensive lowland forest destruction within its range and, perhaps more severely, by hunting and snaring. However, recent evidence suggests that the species may be able to tolerate a higher level of hunting pressure than was previously thought.

25. Species: *Lophura erythropthalma* – Crestless fireback pheasant

Local names: *Singgier* (Bornean), *kuang-bestam* (Malayan).

Description: The males lack a crest; have red facial skin; a rufous tail. This species is the only pheasant species in which the females have a glossy bluish black plumage; head brownish, throat grayish; also have a red orbital patch.

Call: A low repeated '*tak-takrau*' ; '*tooktarod*' ; alarm call '*kak*'.

Habitat: Dense evergreen forest; lowland gently sloping dipterocarp forest.

Food: Insects, berries. Roost on trees at night. Feed in pairs or flocks of 5-6.

IUCN Red List Category: Vulnerable

✠ Major causes: The overriding threats are habitat loss, degradation and fragmentation as a result of large-scale commercial logging, even within some protected areas, and widespread clearance for plantations of rubber and oil-palm. Rates of forest loss in the Sundaic lowlands have been extremely rapid (Kalimantan lost nearly 25% of its evergreen forest during 1985-1997, and Sumatra lost almost 30% of its 1985 cover), because of a variety of factors, including the escalation of illegal logging and land conversion. A significant issue currently is selective logging in an array of forest reserves (not protected areas per se) that results in habitat degradation, including degradation of any unknown features that influence habitat selection by this species, and opens the forest to fire risk, reduced value, and conversion to agriculture in future. Hunting for food may pose an additional, more localised threat.

26. Species: *Lophura ignita* – Crested fireback (Bornean crested fireback pheasant)

Local names: *Ayam pegar* (Malayan), *sempidan* (Bornean).

Description: The male's body plumage is iridescent purplish; brilliant maroon on the lower back and rump; white central tail feathers; blue orbital skin patch; vertical crest. Female cinnamon dorsally, blue orbital skin, short vertical crest.

Call: Long squealing call. Alarm call squirrel-like 'chukun chukan'.

Habitat: Thick evergreen forests near rivers.

Food: Fruits, invertebrates, seeds, leaves. Roost on trees. Forage solitarily in non-breeding season.

IUCN Red List Category: Near Threatened

Major causes: Forest destruction in the Sundaic lowlands of Indonesia and Malaysia has been extensive (Kalimantan lost nearly 25% of its evergreen forest during 1985-1997, and Sumatra lost almost 30% of its 1985 cover), because of a variety of factors, including the escalation of logging and land conversion, with deliberate targeting of all remaining stands of valuable timber including those inside protected areas, plus forest fires (particularly in 1997-1998), and declines are compounded by trapping for the cage-bird industry. However, the species's use of secondary growth implies that it is not immediately threatened.

27. Species: *Lophura bulweri* – Wattled pheasant (Bulwer's pheasant)

Description: The male's body plumage is blackish; a uniquely white fan-like tail; blue facial skin. No crest. Female has dark brown plumage; blue facial skin; no crest.

Call: Alarm call '*kak*'; shrill '*kook!, kook!*'.

Habitat: Evergreen tropical montane, and lowland swamp forests, vicinity of rivers and streams.

Food: Insects, seeds.

IUCN Red List Category: Vulnerable

Major causes: Forest loss, degradation and fragmentation, through large-scale commercial logging, widespread clearance for plantations of rubber and oil-palm, highways, and the extensive recent fires pose the primary threats, compounded more locally by hunting.

Fig. 3.1 A. Blood pheasant female (brown), and male (grey, white and light green). B. A male blood pheasant. Photos by Peteris Erglis, Nepal, 2006.

Fig 3.2 A. A male western tragopan, and B. A female western tragopan in Sarahan Pheasantry, Himachal Pradesh, India. Photos by Gobind Sagar Bhardwaj, 2006.

Fig. 3.3 A male Satyr tragopan in Artis Zoo, Netherlands, by Arjan Haverkamp, 2007.

Fig. 3.4 A female Satyr tragopan in Central Park Zoo, New York. Photo by Staven NG, 2008.

Fig. 3.5 A male Blyth's tragopan in San Diego Zoo, California. Photo by Ashok Khosla, 2007.

Fig. 3.6 A. A male Temminck's tragopan in Columbus Zoo, Ohio, USA. Photo by Ray Anspach, 2008. B. A male Temminck's tragopan in a breeding center for Galliformes in Braga, Portugal. Photo by Hugo Barbosa.

Fig. 3.7 A. A male Cabot's tragopan in Bronx Zoo, New York, USA, 2008. B. A female Cabot's tragopan in Central Park Zoo, New York, USA, 2007. Photos by Staven NG.

Fig. 3.8 A. A male Himalayan monal in Central Park Zoo, New York, USA, 2008.
B. A female Himalayan monal in Central Park Zoo, New York, USA, 2008. Photos
by Staven NG.

Fig. 3.9 A male Chinese monal in Chengdu Wildlife Rescue Center, China, 2007. Photo by Wu Jiawei.

Fig. 3.10 A male Lafayette junglefowl in a breeding center for Galliformes in Braga, Portugal. Photo by Hugo Barbosa.

Fig. 3.11 A. A male black-crested kalij pheasant in Kaziranga National Park, Assam, India. Photo by Aditya Singh, 2009. B. A male white-crested kalij pheasant feeding on a slope covered with balsams in Chhota Shimla, Himachal Pradesh, India. Photo by Anita Chauhan.

Fig. 3.12 A male silver pheasant in Portugal. Photo by Antonio Sardinha, 2007.

Fig. 3.13 A male crestless fireback pheasant in the National Zoo of Malaysia. Photo by Li Xin, May 2007.

Fig. 3.14 A. A close-up of the head of a male Bornean crested fireback pheasant, and B. A male Bornean crested fireback pheasant, in a breeding center for Galliformes in Braga, Potugal. Photos by Hugo Barbosa.

Fig. 3.15 A male blue eared-pheasant in Central Park Zoo, New York, USA. Photo by Staven NG, 2007.

Fig. 3.16 A male Reeves's pheasant in an enclosure in China. Photo by Wu Jiawei.

Fig. 3.17 A close-up of a male golden pheasant in Columbus Zoo Ohio, USA. Photo by Ray Anspach, 2008.

Fig. 3.18 Close-up of a male Lady Amherst's pheasant showing ruffed neck feathers, in Bronx Zoo, New York, USA, 2008. Photo by Staven NG.

Fig 3.19 A male mountain peacock pheasant in a breeding center for Galliformes, Braga, Portugal. Photo by Hugo Barbosa.

Fig. 3.20 A close-up of a male Palawan peacock pheasant in Bronx Zoo, New York, USA, 2008. Photo by Staven NG.

Fig. 3.21 A male Germain's peacock pheasant , and B. A female Germain's peacock pheasant, in a breeding center for Galliformes in Braga, Portugal. Photos by Hugo Barbosa.

Fig. 3.22 A. A close-up of a male Malayan peacock pheasant in Bronx Zoo, and B. A male Malayan peacock pheasant in Bronx Zoo, New York, USA, 2007. Photos by Staven NG.

Fig. 3.23 A close-up of a male Great argus pheasant calling, in the Bronx Zoo, New York, USA. Photo by Staven NG, 2008.

Fig. 3.24 A close-up of a male green peafowl in Bronx Zoo, New York, USA. Photo by Staven NG, 2008.

Fig. 3.25 A male Congo peafowl in the Bronx Zoo, New York, USA. Photo by Staven NG, 2007.

8. Genus *Catreus* – Cheer Pheasant

This genus has one species, *Catreus wallichii*.

The cheer pheasant is a mediun-sized montane pheasant in which sexual dimorphism is slight; and both sexes have an occipital crest and a red orbital patch. The plumage is buffy brown, brown and black, with barring and spotting. The tail is elongated and strongly barred.

Table 3.8 Species statistics at a glance.

Species	Altitude Range	Countries of occurrence of the Species	Population in the Wild	Population Trend
Catreus wallichii (Cheer pheasant)	1220 – 3350 m	India, Nepal, Afghanistan, Pakistan	4,000 - 6,000	decreasing

28. Species: *Catreus wallichi.* – Cheer pheasant

Local names: *Tana* (Shimla district, H.P.), *Cheera, Cherus, Chagan* or *Chura* (Kinnaur district, H.P.), *kahir, chihir* (Nepalese), *tshi-er* (Kumaon).

Description: Short pointed brown crest, red orbital patch present in both sexes, females have a shorter crest and smaller orbital patch. The plumage is buffy brown, brown and black, with barring and spotting. The elongated tail is highly graduated, and strongly barred with buff, black and brown.

Call: Crowing call, '*cher-a-per, cher-a-per, cher, cher, cheria, cheria*', cackling repeated '*waaak*', calling in the morning and evening. Alarm call, '*tuk, tuk*'.

Habitat: Subtropical pine forests to subalpine meadow zones, steep grass-covered hillsides having scattered trees; occur in small flocks; feed in the mornings and evenings; remain under cover during afternoons; at night, roost on low trees, bushes, on rocks, or on the ground.

Food: Berries, insects, seeds, worms, tubers.

IUCN Red List Category: Vulnerable

⊠ Major causes: Having been widely shot for sport in the early 20th century, it is still hunted for food, and its eggs collected for local consumption. The patchy nature of its specialised habitat may render the smallest isolated populations vulnerable to extinctions, and higher levels of disturbance, grazing and the felling of wooded ravines now pose a substantial threat. In particular, hunting pressure and habitat destruction by fire and overgrazing have been implicated in its decline in Pakistan, however, these human controlled practices are also important in maintaining moderately disturbed habitats which the species favours. Conversion of grassland to permanent arable terraces is reducing available habitat, as are schemes to plant mid-altitude grasslands with forest. A key requirement for the long-term survival of this species will be to strike a balance between maintaining the low intensity human practices on which the species relies, while limiting hunting pressure.

9. Genus *Syrmaticus* – The Long-tailed pheasants

This genus has five species - *Syrmaticus reevesi, S. humiae, S.ellioti, S. mikado,* and *S. soemmerringi.*

The long-tailed pheasants are montane pheasants. The sexes are highly dimorphic (distinguishable at sight as male or female). The tail is greatly elongated and strongly barred. Both sexes lack crests. The orbital wattles, if present, are red. Females are strongly marked with brown, black and white; sides and flanks extensively white.

Table 3.9 Species statistics at a glance.

Species	Altitude Range	Countries of occurrence of the Species	Population in the Wild	Population Trend
Syrmaticus reevesi (Reeve's pheasant)	300-3200 m	China, Mongolia	2,500-9,999	decreasing
Syrmaticus humiae (Bar-tailed pheasant)	1200-2400 m	India, Burma, China, Thailand	10,000-19,999	decreasing
Syrmaticus ellioti (Elliot's pheasant)	200-1900 m	China	100,000-499,999	decreasing
Syrmaticus mikado (Mikado pheasant)	1600-3300 m	Taiwan	10,000-19,999	decreasing
Syrmaticus soemmerringi (Copper pheasant)	Upto 1370 m	Japan	unknown	decreasing

29.   Species: *Syrmaticus reevesi* – Reeve's pheasant

Local names: *Chi-ky* (Chinese)

Description: Males have very long, barred tail (upto 62 inches); black and white head. Females have a shorter tail (upto 17 inches), buffy stripes above the eye; buffy throat and neck.

Call: A twittering chuckle; a series of high-pitched chirping call.

Habitat: Wooded and hilly areas of central China; open woodlands of evergreen or deciduous trees; valleys having steep sided canyons with coniferous trees. Forages in morning and late afternoon; although gregarious during the winters, in spring the flocks disperse and form pairs or groups of 1 male and 2 females.

Food: Berries, insects, legumes, acorns, roots, tubers.

IUCN Red List Category: Vulnerable

Major causes: The main threat to the species is the continuing deforestation within its range, which is reducing and fragmenting its habitat. Hunting for food is believed to be an important threat, and its eggs are collected. It was hunted in the past for its long tail feathers, which were used as a decoration in the Peking opera costumes, but plastic feathers are increasingly being used for this purpose.

30. Species: *Syrmaticus humiae* – Bar-tailed pheasant

Local names: *Wit* (Burma), *loe-nin-koi* (Manipur).

Description: Males have long tail (15 inches); strongly barred with gray, brown and black, white stripes on the wings, dark-colored neck. Females have a shorter white-tipped tail barred with black, underparts are barred with brown, brown neck-front.

Call: Display call a loud '*chuck*'; a loud hissing call; clucking. Alarm call repeated '*buk*', and a screeching alarm note.

Habitat: Open oak and pine forests on rocky substrates with light undergrowth. Grass covered hillsides. Feed in the vicinity of streams. Foraging at dawn and dusk, spends afternoon in heavy woods. Occur in small groups of 3-5; roost in the forest cover.

Food: Berries, snails, chestnuts.

IUCN Red List Category: Near Threatened

❖ Major causes: The ease with which it can be trapped has been a major cause of its continuing decline across much of its range, including populations within protected areas. However, the persistence of the species in northern India and in Thailand suggests that it is resilient in the face of heavy exploitation. Extensive shifting cultivation and uncontrolled annual burning has resulted in substantial fragmentation and loss of suitable habitat in Myanmar, China and India. In north Thailand, it has suffered from agricultural intensification and habitat fragmentation resulting from development projects, and reforestation of large areas with dense conifer plantations may also pose a threat.

31.   Species: *Syrmaticus ellioti* – Elliot's pheasant

Local names: *Han-ky* (Chinese).

Description: In males, long tail is strongly barred with white and brown; red orbital patch present, sides of the neck are white to grayish white. Under-parts are white. Females have a shorter tail tipped with white, sides of the neck ashy grey; under-parts white; front of the throat is black.

Call: Display call a loud, repeated '*geke*'. A shrill but not loud squeal by both sexes. Low clucks and chuckles.

Habitat: Includes a variety of forest types – Sparse coniferous forest; mixed evergreen and deciduous forests, thick jungle and ravine vegetation, bamboo thickets. Forages in morning and late-afternoon, in pairs or small family groups.

Food: Seeds, berries, leaves, insects.

IUCN Red List Category: Near Threatened

❖ Major causes: Most of the natural forest within its range has been cleared or modified as a result of the demands for agricultural land and timber, but natural forest clearance has been illegal since 1998. Current threats include the burning of forest by man-made hill fires, collection of firewood, and illegal hunting.

32. Species: *Syrmaticus mikado* – Mikado pheasant

Description: Male plumage generally purplish black, very long and white-barred tail (upto 20 inches), red orbital patch present. Females have a shorter tail barred with black, brown and white; plumage olive brown above, mottled with white below.

Call: Breeding season male call shrill drawn-out whistle '*chiri*'. Chuckling sounds.

Habitat: Steep mountainsides, thick forests having a dense under-growth of rhododendrons, bamboo and ferns. Found in small groups. Feed in the morning and evening. Perch on trees during rainfall.

Food: Seeds, fruits, ferns, leaves, insects.

IUCN Red List Category: Near Threatened

✠ Major causes: Heavy hunting pressure was a problem for it in the past, and appears to be returning at some sites, even within protected areas. It has undoubtedly declined because of habitat loss, due to infrastructure development and landslides, and in the long term, sub-populations may become isolated in protected areas.

33. Species: *Syrmaticus soemmerringi* – Copper pheasant

Local names: *Yamadori* (Japanese).

Description: Male plumage coppery colour, long brown tail (upto 38 inches) with narrow black barring. Female has a shorter white-tipped tail with a black band near the tip; plumage cinnamon to rufous.

Call: Vocalization unknown. Wing-whirring display produces a loud drumming noise.

Habitat: Coniferous forests, adjoining mixed forest with dense undergrowth, grassy mountainsides. Roost 5 m above ground on trees. Relatively solitary, sometimes form small flocks.

Food: Acorns, insects, earthworms, crabs.

IUCN Red List Category: Near Threatened

✠ Major causes: The initial, and perhaps the current, decline was due to hunting; however the shooting of females has been illegal since 1976. In addition to the effects of hunting and habitat loss, feral cats and dogs may be causing a reduction in breeding success, and hybridisation (of the 5 subspecies) between wild and captive-reared stock may also be a problem.

10.   Genus *Phasianus* – The True Pheasants

This genus has two sub-species - *Phasianus colchicus* and *Phasianus c. versicolor.*

The true pheasants are temperate, open-country pheasants. Sexual dimorphism is well developed. Tail elongated and strongly barred. Erectile ear-tufts present. Orbital wattles are red with scattered black plumules. Females have hair-like rump feathers. Tail highly graduated. Males are extensively iridescent over most of their plumage (except abdomen, wing and tail). Females are rather uniformly brown. It is the most widespread species of pheasant.

Table 3.10 Species statistics at a glance.

Species	Altitude Range	Countries of occurrence of the Species	Population in the Wild	Population Trend
Phasianus colchicus (Common pheasant)	1500-2600 m	Iran, former USSR, China, Mongolia, Turkey, Korea, Taiwan, Burma [Introduced in Europe, N. America, Hawaii, Japan, Australia, New Zealand]	45,000,000 - 300,000,000	unset

Phasianus c. versicolor (Green / Japanese pheasant)	Up to 1065 m	Japan [also introduced on Hawaiian Islands]	-	-

34.   Species: *Phasianus colchicus* – Common pheasant

Local names: Black-necked pheasant, Mongolian pheasant, Ring-necked pheasant.

Description: Males have a purplish head, neck, with or without a white neck-ring, a maroon breast tending toward orange on the flanks. Elongated barred tail. Females are generally yellowish buff in colour, with a chestnut tinge on upper body; long tail with black and chestnut barring.

Call: Crowing call is a loud, hoarse '*ko-or-ok, korrk-kok*', or '*kok-ok-ok*', '*ko-koro*'. Challenge call is a hoarse '*krrrah*'.

Habitat: Edges of rivers, hilly areas close to cultivated fields having bamboo groves and low pine thickets; near fields in flat and level land. Feed during morning and evening. Rest during the day; roost on trees, brushy areas, reed beds.

Food: Grains, fruits, insects.

IUCN Red List Category: Least Concern

Sub-species: *Phasianus c. versicolor* – Green/ Japanese pheasant

Description: Males have a greenish to blueish iridescent plumage on head, neck, breast and mantle; olive black wings; heavy black barring on the tail. Females are darker than common pheasant, with greenish sheen.

Call: Slightly different from Common pheasant.

Habitat: Sparse woods or brush lands near cultivated fields, grassy areas near rivers, low hilly areas near the coast. Forage in small groups during morning and evening hours near agricultural fields and open grasslands.

Food: Seeds, berries, acorns, chestnuts, snails, crustaceans, fruits, insects.

11.   Genus *Chrysolophus* – The Ruffed Pheasants

This genus has two species - *Chrysolophus pictus* and *C. amherstiae.*

The ruffed pheasants are montane pheasants. Sexual dimorphism is well developed. The males have a short decumbent crest and ornamental ruffs, which are expanded during display. The long tail is strongly graduated. A small orbital skin patch present in males. Females are barred with dark brown and buff.

Table 3.11 Species statistics at a glance.

Species	Altitude Range	Countries of occurrence of the Species	Population in the Wild	Population Trend
Chrysolophus pictus (Golden pheasant)	Below 1980 m	China, Tibet, (introduced in Great Britain)	unknown	unset
Chrysolophus amherstiae (Lady Amherst's pheasant)	2135-4575 m	China, Tibet, Burma, (introduced in Great Britain)	20,000-49,999	unset

35.   Species: *Chrysolophus pictus* – Golden pheasant

Local names: *kin-ky, ching chi* (Chinese).

Description: Male has golden nape, striated pattern on ruff, orange red (or golden yellow) feathers on under-parts and rump. Females yellowish on the crown, slightly barred with blackish throughout; yellow orbital skin.

Call: Crowing call '*chak*' or '*cha-chak*'. Display call loud squeaking and hissing.

Habitat: Mountain slopes and valleys, with a dense growth of bamboos and bushes. Roost on Pine trees 3 m above ground. Found singly, in pairs or trios.

Food: Leaves and tender shoots of shrubs and bamboo, flowers of rhododendrons, insects.

IUCN Red List Category: Least Concern

36.   Species: *Chrysolophus amherstiae* – Lady Amherst's pheasant

Local names: *wokree* (Burma), *seng-ky, kwa-kwa-chi, sun-chi* – fowl of the buds (Chinese).

Description: Male has a long white tail barred with black (up to 37 inches), scaly black and white pattern on ruff. Female orbital skin is light blue.

Call: Crowing call similar to Golden pheasant.

Habitat: Found in high mountain wooded slopes; lives in bamboo cover or dense bushes.

Food: Insects, ferns, bamboo sprouts, grains, nuts, seeds.

IUCN Red List Category: Least Concern

12.   Genus *Polyplectron* – The Peacock Pheasants

This genus has seven species - *Polyplectron chalcurum, P. inopinatum, P. bicalcaratum, P. germaini, P. malacense, P. schleiermacheri, P. emphanum.*

A newly-split species, Hainan peacock pheasant *Polyplectron katsumatae*, is listed as Endangered as it has very small population (in Hainan island, China) which is estimated to have undergone a very rapid decline owing to habitat loss and hunting. Surveys are urgently required to obtain an up-to-date population estimate. Apart from being genetically distinct, it is recognisably different from the Grey Peacock-pheasant *P. bicalcaratum* in having a smaller, forward-pointing and bushy crest, extensive vermiculated grey-brown plumage, with large ocelli on the upperparts, each spot being green or blue with a buffish or bold white surround. The upper throat is whitish whilst the bare facial skin is pinkish or yellowish (Collar, 2009).

The peacock pheasants are tropical pheasants. Sexual dimorphism is moderately developed. The tail is rounded, graduated, with iridescent bluish green ocelli or banding in males. Bare orbital areas are inconspicuous or only moderately developed in males, wattles are lacking, crests are short to moderate in length. Females are duller in color, usually lack iridescence.

Table 3.12 Species statistics at a glance.

Species	Altitude Range	Countries of occurrence of the Species	Population in the Wild	Popln. Trend
Polyplectron chalcurum (Bronze-tailed/Sumatran pheasant)	150-1220 m	Sumatra	unknown	unset
Polyplectron inopinatum Mountain/ Rothschild peacock pheasant	900-1800 m	Malaysia	2,500-9,999	decreasing

Polyplectron bicalcaratum (Grey peacock pheasant)	0-1830 m	India, Burma, China, Bhutan, Thailand, Laos, Tibet, Vietnam	unknown	unset
Polyplectron germaini (Germain's peacock pheasant)	0-1220 m	Vietnam	10,000-19,999	decreasing
Polyplectron malacense (Malayan peacock pheasant)	0-305 m	Malaysia	2,500-9,999	decreasing
Polyplectron schleiermacheri (Bornean peacock pheasant)		Borneo	1,000-2,499	decreasing
Polyplectron emphanum (Palawan peacock pheasant)	0-660 m	Palawan island	2,500-9,999	decreasing

37. Species: *Polyplectron chalcurum* – Bronze-tailed/Sumatran peacock pheasant

Local names: *Karo-karo, loekei* (Sumatran)

Description: Chestnut brown plumage, which lacks definite ocelli, pointed tail which lacks ocelli, iridescence limited to some barring and patches. In both sexes the facial area is almost completely feathered, no orbital skin patch.

Call: A twittering and repeated '*pitt*' call.

Habitat: Found in thick montane forests of Sumatra. Quiet and secretive birds. Mostly under cover. Occur in pairs or family groups.

Food: Fruits, insects, worms.

IUCN Red List Category: Least Concern

38. Species: *Polyplectron inopinatum* – Mountain/Rothschild peacock pheasant

Local names: Mirror pheasant

Description: Male lacks crest, face fully feathered. The face and head are grayer than the rest of the upper plumage which tends to be chestnut. In both sexes the upper parts are dotted with small, dark ocelli, large ocelli on lateral tail feathers.

Call: Conversational 'chack' call; a quiet descending whistle.

Habitat: Rugged and wooded mountain habitats. Feed in groups.

Food: Fruits, insects, worms, petals.

IUCN Red List Category: Vulnerable

▨ Major causes: Conversion of forest for agriculture around its lower altitudinal limits may be causing some declines. There is a considerable danger that a proposed north-south road linking the hill stations of Genting Highlands, Fraser's Hill and Cameron Highlands will result in the further fragmentation and degradation of a substantial area of its montane habitat. These plans have been shelved, but should they be raised again, the species may warrant uplisting.

39. Species: *Polyplectron bicalcaratum* – Grey peacock pheasant

Local names: *Chinquis* (Chinese).

Description: Males grayish, females brownish, whitish throats, dark ocelli on the back and wings. The male has a bushy crest that is erected anteriorly in front of the eyes. The tail is rounded, with large iridescent ocelli.

Call: Whistled '*trew-tree*' or '*taa-pwi*'. Harsh repeated '*putta*'. Crowing call repeated '*phee-hoo*'. A series of croaking calls '*wak-wak-wak, qua-qua-qua*' and '*ok-kok-kok*'.

Habitat: Dense evergreen and semi-evergreen forests. Tropical lowland forests up to 1830 m. Thick cover along streamside banks and in dense evergreen forests, tangled scrub, or mixed bamboo and thick scrub. Secretive, slow, methodical; occur in pairs or small groups.

Food: Fruits, insects, grain, berries, seeds, wild figs, snails.

IUCN Red List Category: Least Concern

40. Species: *Polyplectron germaini* – Germain's peacock pheasant

Description: Bare facial skin is distinctly reddish, short bushy crest, rounded tail with numerous ocelli. Brown to brownish gray plumage.

Call: High-pitched 'hwo-hwoit' repeated 4-6 times. Alarm call loud, fast cackle.

Habitat: Humid forests. Mostly solitary, also found in pairs or family groups.

Food: Fruits, insects, grain, berries, seeds, wild figs, snails.

IUCN Red List Category: Near Threatened

Major causes: Historically it has suffered major declines due to forest loss and fragmentation resulting from commercial logging and resettlement programmes. These activities have exacerbated clearance of land for subsistence cultivation and localised commercial cropping of coffee and cashew nuts. However, in recent times habitat loss has slowed in Indochina. Hunting with guns and snares, even within protected areas, also represents a threat to the species. Shortage of staff and resources in protected areas results in ineffective control of illegal activities, especially hunting, disturbance and small-scale logging.

41.   Species: *Polyplectron malacense* – Malayan peacock pheasant

Local names: *kuan, kuang* (Malayan).

Description: Males have orange facial skin, long greenish blue crest posteriorly oriented, brownish plumage studded with many iridescent ocelli on the back and tail. Female uniformly brown, pale gray throat.

Call: Loud cackling '*kwok kwok*'.

Habitat: Lowland foothill forests of interior Malaysia. Tall lowland Dipterocarp forests, gently sloping ground, dense undergrowth near water. Usually occur singly or in pairs.

Food: Fruits, insects, berries.

IUCN Red List Category: Vulnerable

Major causes: Lowland forest clearance and modification for cultivation remain the major threats. Only 25% of suitable habitat that was available for the species prior to 1970 remains today. Hunting for food, sport and the bird trade presumably contributed to its probable extinction in Thailand. Whilst it is susceptible to snaring targeted at all ground foraging animals, there is no evidence to indicate it is particularly sought after in Malaysia.

42.   Species: *Polyplectron schleiermacheri* – Bornean peacock pheasant

Description: Males have nearly black underparts, iridescent green on the sides of the neck and forebreast, a crest which is greenish to violet with recurved feathers.

Habitat: Lowland forests.

Food: Fruits, insects, berries.

IUCN Red List Category: Endangered

✥ Major causes: In central Kalimantan, habitat loss, degradation and fragmentation as a result of large-scale commercial logging (deliberately targeting all remaining stands of valuable timber including those inside protected areas, widespread clearance for plantations of rubber and oil-palm, and hunting with snares, are the main threats. Kalimantan lost nearly 25% of its evergreen forest during 1985-1997. The impact of the major fires of 1997-1998 was patchy, with many small alluvial areas escaping damage. However, such fires appear to be increasing in frequency and severity. In central Kalimantan, most remaining lowland forest is granted to logging concessions, with a negligible area currently afforded any protected status. The species was recorded in trade by TRAFFIC in 1998 when six birds were taken out of Kalimantan to Singapore.

43.   Species: *Polyplectron emphanum* – Palawan peacock pheasant

Local names: *sulu maläk, dusan bertik* (Palawan).

Description: Males have extensive amounts of green iridescence on the neck, mantle, and inner wings, long erectile crest. The facial skin around the eyes is bright red. Tail has a buffy terminal band. Females have ocelli on outer tail feathers; generally dark brownish black, whitish gray throat, pale gray above the eyes.

Habitat: Damp forests of coastal plains, scrub of foothills. Occur in pairs.

Food: Fruits, insects, seeds.

IUCN Red List Category: Vulnerable

✥ Major causes: Deforestation in lowland Palawan has been extensive, and logging and mining concessions have been granted for almost all remaining forest on the island. Illegal logging is thought to persist in the remaining extensive forest of the south. Forest at Iwahig Penal Colony, regarded as a key site, may be threatened by plans to mine chromite. By the late 1960s, it was being extensively hunted and trapped in large numbers for live trade, but exports were much reduced by the late 1980s. In the mid-1990s, it was heavily hunted adjacent to Puerto Princesa Subterranean River National Park.

13. Genus *Rheinardia* – The Crested Argus Pheasant

This genus has one species - *Rheinardia ocellata.*

The crested argus is a large tropical pheasant. Sexual dimorphism is well developed. The males have a very elongated tail, which is strongly graduated. Both sexes have a short occipital crest. Most of the plumage is uniformly brown with blackish barring and buffy freckles or vermiculations. Females have a shorter less ornamental tail.

Table 3.13 Species statistics at a glance.

Species	Altitude Range	Countries of occurrence of the Species	Population in the Wild	Population Trend
Rheinardia ocellata (Crested Argus)	0-1525 m	Vietnam, Laos, Malay	10,000-19,999	decreasing

44. Species: *Rheinardia ocellata* – Crested argus pheasant

Local names: *tri* (Vietnamese).

Description: Males have a short bi-colored crest. Central tail feathers elongated (67 inches). Females also crested, pale grayish face, dark brown neck and body plumage. Somewhat barred pattern on wing and tail.

Call: Loud '*kuau*' or '*ho-huiho*' repeated several times. Breeding season '*oo-kia-wau*'. Soft, chuckling notes by both sexes.

Habitat: Heavy forests; hilly lowland and lower montane forests, on very damp slopes of mountains. Rare and elusive. Mostly solitary.

Food: Fruits, insects, frogs, ferns.

IUCN Red List Category: Near Threatened

✠ Major causes: The Indo-Chinese population is probably most at risk from continuing forest loss and degradation, both within and outside protected areas. The greatest problems stem from commercial logging, illegal timber extraction, clearance for agricultural plantations, encroachment by shifting cultivators, and road-building. Disturbance and snaring at display arenas are more significant threats than deforestation in some areas. The Malaysian population is less threatened, with the main documented threat being limited habitat loss on the periphery of Taman Negara, although its narrow altitudinal range in this country lies mostly outside protected areas, exposing it to disturbance from logging.

14.   Genus *Argusianus* – The Great Argus Pheasant

This genus has one species - *Argusianus argus.*

The great argus is a very large tropical pheasant. Sexual dimorphism is highly developed. Iridescence is limited to the ocelli on the wings of the male. The central pair of tail feathers of males are extremely elongated; short occipital crest. Female's skin on face and neck is bluish. The body feathers are mostly spotted and finely barred with brown, buff and chestnut.

The genus also has an extinct species of argus, called the double-banded argus. The mysterious pheasant, the double-banded argus, *Argusianus bipunctatus*, is known only from the portion of a single primary flight feather. It was described in 1871 from this feather piece, found in a millinery shipment (for the designing and making of hats) imported to London. Its origin was hypothesized to be from Java, Indonesia, or Tioman Island of Malaysia, because of the great argus' absence from these locations. The IUCN, following the precautionary principle, lists this taxon as 'extinct' due to loss of its habitat. While the feather is indeed quite distinct, it represents a fairly simple divergence: the entirely asymmetrically-patterned vanes of the feather as in great argus are instead near-symmetrical, and both sides bear the darker brown shaftward area with innumerable whitish speckles. The feather is now housed in the British Natural History Museum (Wikipedia). Although most scientists now regard it as a synonym of *A. argus* (mainly due to the lack of DNA evidence) (Davison and McGowan, 2009), my opinion is that *A. bipunctatus* is an extinct sub-species or geographical race that evolved in the same way as other island pheasants, such as Japanese pheasant and Hainan peacock pheasant.

Table 3.14 Species statistics at a glance.

Species	Altitude Range	Countries of occurrence of the Species	Population in the Wild	Population Trend
Argusianus argus (Great Argus)	0-915 m	Borneo, Sumatra, Malaysia	100,000	decreasing

45. Species: *Argusianus argus* - Great argus pheasant

Local names: *keee* (Dutch Bornean), *Kuang raya* (North Malayan), *koeweau* (Sumatran).

Description: Males have an enormously long and narrow central tail feathers (42 inches). Both sexes have only short crests and reddish feet.

Call: The male's long call is a loud, musical '*kwow-wow*' or '*kweau*'. Female has repeated '*wow*' notes.

Habitat: Heavy tropical forest; lowland and hill Dipterocarp forest. Feed in the morning and evening; perch on trees at afternoon and night; relatively solitary, also occur in family groups.

Food: Fruits, insects, mollusks, leaves, nuts, seeds.

IUCN Red List Category: Near Threatened

Major causes: Forest destruction in the Sundaic lowlands of Indonesia and Malaysia has been extensive (Kalimantan lost nearly 25% of its evergreen forest during 1985-1997, and Sumatra lost almost 30% of its 1985 cover), due to escalation of logging and land conversion, with deliberate targeting of all remaining stands of valuable timber including those inside protected areas, plus forest fires, and declines are compounded by trapping for the cage-bird industry.

However, the species's use of secondary growth and higher elevations implies that it is not immediately threatened.

15. Genus *Pavo* – The Peafowl

This genus has two species - *Pavo cristatus* and *P. muticus.*

The peafowl are very large tropical pheasants. Sexual dimorphism is highly developed. Both sexes exhibit iridescent plumage. Males have a bright orbital skin patch. In males, the tail is hidden by very elongated and ornamental tail-feathers, which are iridescent and tipped with complex ocelli. In females the plumage is generally less iridescent.

Table 3.15 Species statistics at a glance.

Species	Altitude Range	Countries of occurrence of the Species	Population in the Wild	Population Trend
Pavo cristatus (Indian peacock)	Upto 1830 m	India, Sri Lanka, Pakistan, Bangladesh, Nepal	unknown	unset
Pavo cristatus (Green peafowl)	Plains up to 915 m	India, Burma, Thailand, China, Java, Bangladesh	10,000-19,999	decreasing

46. Species: *Pavo cristatus* – Indian peafowl

Local names: *mor* (Hindi), *monara* (Ceylonese).

Description: Males and females have whitish cheeks and a tufted, fan-like crest. Females lack the long and iridescent tail-feathers; their neck and head is iridescent.

Call: Crowing '*he-on*' or '*kee-ow*', loud trumpeting '*may-awe*'. Breeding '*kayong-kayong-kayong*'.

Habitat: Moist or dry deciduous forests near streams, with bushy undergrowth. Usually forage in small groups; roost on trees; rest and preen in shade near water during daytime.

Food: Insects, small reptiles, mammals, berries, wild figs.

IUCN Red List Category: Least Concern

47. Species: *Pavo muticus* – Green peafowl

Local names: *burong merak* (Malayan), *oodoung* (Burma).

Description: Males and females have a long, tapering, narrow crest, bluish and yellowish facial skin. In females, under-parts tend towards blackish.

Call: Crowing '*waaaa-ak*', '*aow-aow*'. Alarm note '*kwok*'. Guttural clucking sounds.

Habitat: Elephant grass, open dry forest, evergreen forest with dense bush undergrowth. Come to the river in the morning to drink and preen. Feed in the morning. Return to the river in the evening. Rest during the afternoon. Roost on trees at night.

Food: Insects, seeds, small mammals, berries, flower petals.

IUCN Red List Category: Endangered

❇ Major causes: Widespread hunting for meat and feathers, and collection of eggs and chicks, combined with habitat modification and human disturbance, has caused this species's catastrophic decline. Fragmentation has isolated many small populations, increasing their susceptibility to local extinction, but selective logging appears to have no adverse effects on peafowl distribution. Other threats may include trade in the male's spectacular train feathers. It is regarded as a crop-pest by farmers in China and Thailand, and is consequently poisoned. The spread of human settlement presents the greatest threat; directly through hunting pressure and habitat loss but also indirectly by preventing access to otherwise suitable habitat.

16. Genus *Afropavo* – The Congo Peacock

This genus has one species - *Afropavo congensis.*

The Congo peacock is a tropical pheasant. Sexual dimorphism is well developed. Both the sexes have iridescent plumage and crests. A large area of bluish to reddish skin is present on the sides of the face and upper neck. In males there is a tuft of white bristles in addition to softer black crest feathers.

Table 3.16 Species statistics at a glance.

Species	Altitude Range	Countries of occurrence of the Species	Population in the Wild	Population Trend
Afropavo congensis (Congo peacock)	100- 1200 m	Congo	2,500- 9,999	decreasing

48. Species: *Afropavo congensis* – Congo peacock

Description: Shy peafowl with green upperparts. Male has dark bronze-green upperparts and black underparts, short black and dense white, bristly crown, naked red throat. Violet-blue wing-coverts, breast feathers, and end of tail feathers. Lead-grey bill and grey feet. Long spur on each leg. Female slightly smaller, rusty-brown with glossy green upperparts, and short brown crown. Males lack a train of tail feathers, have no ocelli.

Call: Crowing '*rro-ho-ho-o-a*', '*gowe-gowah*', *ko-ko-wa*. Female '*hej-hoh hej-hoh*'.

Habitat: Rainforest of the east-central Congo Basin. Occur in pairs.

Food: Insects, seeds, fruits.

IUCN Red List Category: Vulnerable

✠ Major causes: Historically, its population was probably reduced by forest clearance and hunting. Presently, habitat is being lost to mining, subsistence agriculture and logging at several locations. Mining results in the opening up of remote areas, with a corresponding increase in hunting. Surveys have revealed high hunting pressure in the Kisangani region and around Salonga National Park. The capture rate of the species for each village in and around the Salonga National Park is around 20 birds per year. In addition, the species's eggs are collected. The presence of guerrilla fighters and huge numbers of Rwandan refugees in the eastern DRC since 1994 also poses a significant threat because of increased hunting and habitat loss.

Sources: IUCN Red List, BirdLife International.

✠ Source: BirdLife International - Species fact-sheet. URL: http://www.birdlife.org

Chapter 4

How Conservation Science Helps

It has been more than a hundred years since Dr. Beebe went on the expedition to Asia, to study and collect specimens for writing the monograph describing the then little-known pheasant species. Since then, Asia, and Southeast Asia in particular, has developed vast urban areas and commercial tree plantations. Most of this development has taken place by encroaching upon forest land that was home to a rich diversity of wildlife, including the pheasants. The decreasing pheasant habitat and population in the past few decades have prompted conservation efforts by international organizations, resulting in comprehensive legislations and important steps for conservation.

Fig. 4 A painting of a pair of koklass pheasants by H C Richter in *The Birds of Asia*, vol. 7 by John Gould.

Chapter 4

How Conservation Science Helps

Birds are a considerable component of our natural heritage. They are valued for economic, cultural, and spiritual reasons. They inspire every culture, every religion, and every nation in some way, through their beauty and calls, and their close links with people. They help to define national and local identities in the face of globalization of societies (BirdLife International, 2001). Over the centuries, mythology, folklore, art and sculpture, drama, song and dance, literature, postal stamps and currency notes, and just about any other manifestation of culture has been inspired by birds (Kothari, 2007); and bird images are often used to adorn household articles.

In nature in general, places that are rich in bird species are also rich in other animal species, and thus birds are an indicator that can be used to locate species-rich areas, and to detect environmental changes and problems. For instance, the loss of threatened birds from various parts of Asia is an indicator of a widespread loss of other biodiversity, and a deteriorating natural environment. Also, birds are high up in the food chain and hence they reflect changes at lower levels (BirdLife International, 2001).

A wealth of data on birds has been collected by ornithologists, and their conservation status is better known compared to other groups of animals. Among birds, 'the Galliformes are one of the most threatened of taxonomic orders, in terms of the proportion of species threatened. A part of the reason for this is their close relationship with man. While the three species which are most celebrated for their association with man, the red junglefowl *Gallus gallus*, the ring-necked pheasant *Phasianus colchicus*, and the Indian peafowl *Pavo muticus*, are not globally at risk, many other species are threatened with extinction as a result of human interference through hunting and habitat loss' (McGowan, 1994).

Information on threatened bird species in Asia still needs to be greatly improved for the purpose of conservation. 80% of all threatened bird species in Asia require baseline surveys to map their current distributions, and to estimate their population sizes more accurately. In addition, >50% require research into their ecology, and >20% require research to determine the nature and causes of threats impacting them. For only 20% of species is the current information sufficient enough to provide a baseline for monitoring changes in status (BirdLife International, 2001).

Furthermore, suitable social and economic policies are required for conservation success. For example, improved standards of living, conservation awareness amongst local people of all social strata, better training and infrastructure in forestry departments, are all fundamental to the success of conservation actions. Conservation efforts need to be designed in ways which, where possible, provide benefits or compensation for local people and economies (BirdLife International, 2001).

For this purpose, national legislations and international agreements are important and provide a framework for conservation. Consequently, a range of national, regional, and international agreements have been established to support the conservation of species and ecosystems, and the sustainability of the use of natural resources. Such agreements include the Convention on Biological Diversity (CBD), the Bonn Convention on the Conservation of Migratory Species, the Ramsar Convention on Wetlands of International Importance, the Convention on International Trade in Endangered Species of wild fauna and flora (CITES), the International Tropical Timber Agreement, the South Asia Cooperative Environment Programme, and the ASEAN Agreement on the Conservation of Nature and Natural Resources, as well as numerous national initiatives (BirdLife International, 2001).

Apart from the wildlife laws and environmental laws in support of these agreements, other types of national legislations and policies can have significant impacts on bird species, particularly those on forestry, agriculture, sustainability, landuse planning, land ownership, and environmental impact assessment. By ensuring that laws (at local, state/provincial, and national scales) take into account the principle of conservation, threats to species can be minimized (BirdLife International, 2001).

Information on threatened species can make a substantial input into the National Biodiversity Strategy and Action Plans (NBSAPs), being developed by parties to the Convention on Biological Diversity, which has been ratified by 21 countries in the region. Article 6 of the CBD binds parties, which includes almost every country in Southeast Asia, to 'develop national strategies, plans, or programmes for the conservation and sustainable use of biological diversity'. One conservation model proposed is the 'Important Bird Area' (IBA) programme of BirdLife International, an IUCN partner organization. This programme is a logical extension of the Red Data Book programme, and targets the identification and protection of key areas, not only containing: (a) globally threatened species, but also, (b) restricted-range species, (c) species which are vulnerable through their colonial or congregatory behaviour at some stage in their life cycle (e.g. seabirds), and (d) species which are endemic to particular biomes. IBAs are thus designed to protect a wide range of bird species through these four target categories. BirdLife International has also been investing in the development of information management tools to support the activities of the partnership, starting with the Important Bird Areas module in 1994. In 1998, the database was revised and extended so that it now covers sites, species, and Endemic Bird Areas. Much of these data are available through the 'Data Zone'. With information on some 10,000 species of bird, over 8,000 IBAs, and 218 EBAs, and together with spatial data, multimedia files, other documents and links, the BirdLife Data Zone is a valuable information resource on the internet (BirdLife International, 2001).

Conserving birds clearly has great ecological and economic benefits. As economies develop, large middle classes are emerging in Asian cities, people with leisure time for bird-watching and other pastimes that re-connect them to the natural environment. Through environmental NGOs, they are growing into a strong lobby for sustainable development, to balance economic growth with the need to protect habitats and wildlife. Ever increasing numbers of people are joining bird societies (the Wild Bird Society of Japan (BirdLife International in Japan) has around 50,000 members); and similar organizations elsewhere in Asia are growing swiftly (BirdLife International, 2001).

The bird-watching industry is becoming a major economic force for the tourism and recreation industries. In the past 2 decades or so, a lot of interest in bird-watching activities has been generated in many Asian and African countries. Protected areas and other wildlife tourism sites are attracting international as well as domestic tourists. Guided bird-watching tours are popular in countries having a rich avifauna - China, Japan, Thailand, Malaysia, Indonesia, Philippines, and also African countries. Forest species like pheasants, hornbills, pittas, trogons, bristleheads, and waterfowl species like ducks, ibises, storks, sandpipers, cranes, are benefitting from *and* helping to develop tourism infrastructure – hotels, restaurants, guides, and transportation. A number of books and field-identification guides for birds of each country are now available. 'Twitching', a form of bird-watching that originated in Europe, is now an interesting global phenomenon, and books written by 'twitchers' are further popularizing bird-watching among the general public. There are also a number of websites and blogs where bird-watchers share their photographs and stories of travel adventures. One such popular site is Fatbirder.com. If managed properly, wildlife tourism provides a sustainable livelihood for local people, contributes to regional economy, helps to increase conservation awareness, and helps to generate funds for conservation projects.

International Organisations

In order to understand the conservation scenario, we should be aware about the main international organizations working to save the pheasants, and how the conservation plans and action are carried out. Typically, conservation organizations, universities and specialized institutes carry out the research related to the birds and their habitat, and the government departments are responsible for the management of habitats, and formulating policy to support the conservation programs. Specialized Non Government Organisations (NGOs) provide awareness, education, and social support to the local communities affected by conservation programs, and garner support through awareness programs for the general public.

The IUCN

The international organization active in conservation of biodiversity, IUCN, the International Union for Conservation of Nature, is the world's oldest and largest global environmental network of more than one thousand government and NGO member organizations, and has almost 11000 volunteer scientists in more than 160 countries. IUCN has 60 offices around the world. The Head Quarters of IUCN are located in Gland, Switzerland. The IUCN was founded in 1948 as the world's first global environmental organization following an International conference in Fontainebleau, France. It was then known as the International Union for the Protection of Nature (IUPN).

For more than the past 60 years, the IUCN has lead development of conservation science and knowledge, and brought together governments, NGOs, scientists, companies and community organizations to help the countries of the world make better conservation and development decisions. Their mission is "to influence, encourage and assist societies throughout the world to conserve the integrity and diversity of nature and to ensure that any use of natural resources is equitable and ecologically sustainable". IUCN also provides the World Heritage Committee (of UNESCO) with technical evaluations of natural heritage properties, and through its worldwide network of specialists, reports on the state of conservation of listed properties.

IUCN'S Commissions

The 11000 volunteer scientists and experts working for IUCN are grouped into six Commissions. The six Commissions unite the experts from a variety of disciplines, who assess the state of the world's natural resources and provide the Union with sound knowhow and policy advice on conservations issues. The six Commissions are:

Commission on Education & Communication (CEC), Commission on Environmental, Economic and Social Policy (CEESP), Commission on Environmental Law (CEL), Commission on Ecosystem Management (CEM), Species Survival Commission (SSC), World Commission on Protected Areas (WCPA).

Species Survival Commission

The Species Survival Commission is a science-based network of some 7000 volunteer experts from almost every country of the world, all working towards achieving the vision of "a world that values and conserves present levels of biodiversity". These experts are grouped in more than 100 Specialist Groups and Task Forces. Some groups address conservation issues related to particular groups of plants or animals while others focus on topical issues, such as re-introduction of species into former habitat, or wildlife health. Members include –

1.   Researchers
2.   Government officials
3.   Wildlife vets
4.   Zoo and Botanical Institutes employees
5.   Marine Biologists
6.   Protected Area Managers
7.   Experts on plants, birds, mammals, fish, amphibian, reptiles and invertebrates.

Fig. 4.1 The 'Status Survey and Conservation Action Plan for Pheasants' by IUCN, 2000-2004.

The Species Survival Commission of IUCN has promoted the production of Action Plans (Five Year Plan of Action) since 1986. The aim of the Action Plans is to assess the nature and scale of threat to particular species, and to propose conservation action that will lead to a safer future for the species of concern. 57 Action Plans have been produced so far for plant and animal groups. The Action Plans are compiled by SSC's

Specialist Groups. Apart from publishing the Action Plans, the SSC also publishes the 'IUCN Red List of Threatened Animals & Plants'. Red List is a list of all animal and plant species and sub-species that have been assessed according to the IUCN Red List Categories and Criteria.

One of the Specialist Groups, Pheasant Specialist Group, has produced a Conservation Action Plan for pheasants (2000-2004). One of the organizations associated with PSG is the World Pheasant Association. The World Pheasant Association is a registered charity founded in 1975 which aims to develop and promote the conservation of all the species within the order Galliformes, known as the game-birds of the world. In the year 2011, PSG has been merged into the Galliform Specialist Group, which is now responsible for producing the action plans for pheasants as well as cracids, megapodes, grouse, and partridges, quails, and francolins.

The IUCN Red List – Why and How a Species is Listed as Endangered

Many plant and animal species populations have dwindled to critical levels, important habitats are being destroyed, fragmented, and degraded, and ecosystems are being destabilized through climate change, pollution, invasive species, and direct human impacts. At the same time, there is also growing awareness about how biodiversity supports livelihoods, allows sustainable development, and fosters cooperation between nations. This awareness is generated through products such as the IUCN Red List of Threatened Species.

The IUCN Red List is the world's most comprehensive inventory of the global conservation status of plant and animal species. It uses a set of criteria to evaluate the extinction risk of thousands of species and subspecies. This set of criteria is relevant to all the species and all regions of the world. Species in need of conservation attention due to decrease in their population numbers are identified, along with the factors causing their threatened status. The Red List is recognized as the most authoritative guide to the status of biological diversity. More details are available on the website www.iucnredlist.org. The Red List is updated every year.

The IUCN Red List Partnership consists of members and partners who are making a particularly significant contribution to The IUCN Red List of Threatened Species. The partnership includes: Conservation International, BirdLife International, NatureServe, and Zoological Society of London.

How is the Red List compiled?

The IUCN Red List – Categories and Criteria are intended to be an easily understood system for classifying species at high risk of global extinction. There are 9 Categories in the IUCN Red List System (IUCN, 2001):

Extinct (EX)

Extinct in the Wild (EW)

Critically Endangered (CR)

Endangered (EN)

Vulnerable (VU)

Near Threatened (NT)

Least Concern (LC)

Data Deficient (DD)

Not Evaluated (NE)

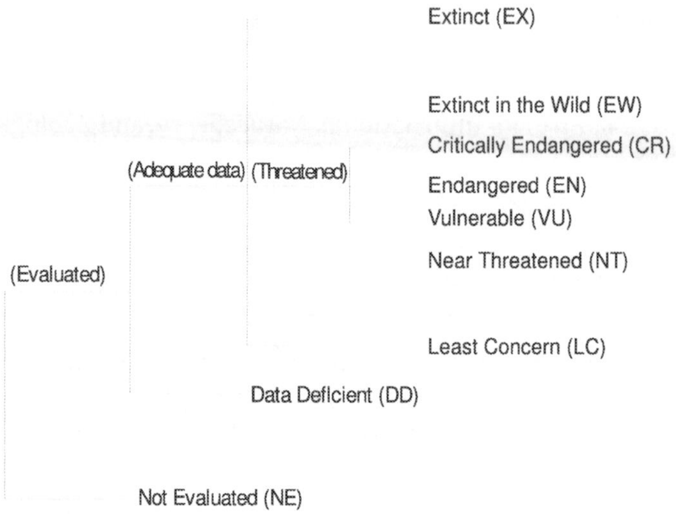

Extinct (EX)

Extinct in the Wild (EW)

Critically Endangered (CR)

(Adequate data) (Threatened)

Endangered (EN)

Vulnerable (VU)

Near Threatened (NT)

(Evaluated)

Least Concern (LC)

Data Deficient (DD)

Not Evaluated (NE)

Fig. 4.2 The 9 categories in the IUCN Red List.

The Categories and their Applications

Out of these categories, the classification of species into Threatened Categories (for species threatened with extinction) - CR, EN and VU, is done using a set of 5 quantitative criteria each. These criteria are for an assessment of the extinction risk and include: rate of decline, population size, area of geographic distribution, and degree of population fragmentation (IUCN, 2001).

The IUCN has adopted the latest version of IUCN Red List Categories and Criteria Version 3.1. These criteria have gone through changes after first being introduced in 1991 by Mace and Lande (Version 1.0). Since then they have been fine-tuned in Version 2.0(1992), 2.1(1993), 2.2(1994), 2.3(1994), 3.0(1999), and finally Version 3.1 (2001) (IUCN, 2001).

The 3 sets of 5 criteria each for 'Critically Endangered, Endangered, and Vulnerable' species are also listed in the IUCN website. To illustrate how species are considered as belonging to one of these categories, 'Criteria for Critically Endangered' species are mentioned in Appendix 2.

Extinct (EX)

A taxon (a unit of classification, eg. species) is Extinct when there is no reasonable doubt that the last individual has died. A taxon is presumed Extinct when exhaustive surveys in known and/or expected habitat, at appropriate times (diurnal, seasonal, annual) throughout its historic range have failed to record an individual. Surveys should be over a time frame appropriate to the taxon's life cycle and life form (IUCN, 2001).

Extinct in the Wild (EW)

A taxon is Extinct in the Wild when it is known only to survive in cultivation, in captivity, or as a naturalized population(s) well outside the historical range. A taxon is presumed Extinct in the Wild when exhaustive surveys in known and/or expected habitat, at appropriate times (diurnal, seasonal, annual), throughout its historic range have failed to record an individual. Surveys should be over a time frame appropriate to the taxon's life cycle and life form (IUCN, 2001).

Critically Endangered (CR)

A taxon is Critically Endangered when the best available evidence indicates that it meets any of the criteria A to E for Critically Endangered, and it is therefore considered to be facing an extremely high risk of extinction in the wild (IUCN, 2001).

Endangered (EN)

A taxon is Endangered when the best available evidence indicates that it meets any of the criteria A to E for Endangered, and it is therefore considered to be facing a very high risk of extinction in the wild (IUCN, 2001).

Vulnerable (VU)

A taxon is Vulnerable when the best available evidence indicates that it meets any of the criteria A to E for Vulnerable, and it is therefore considered to be facing a high risk of extinction in the wild (IUCN, 2001).

Near Threatened (NT)

A taxon is Near Threatened when it has been evaluated against the criteria, and does not qualify for CR, EN or VU now; but it is close to qualifying for, or is likely to qualify for, a threatened category in the near future (IUCN, 2001).

Least Concern (LC)

A taxon is Least Concern when it has been evaluated against the criteria and does not qualify for CR, EN, VU or NT. Widespread and abundant taxa are included in this category (IUCN, 2001).

Data Deficient (DD)

A taxon is Data Deficient when there is inadequate information to make a direct, or indirect assessment of its risk of extinction based on its distribution and/or population status. A taxon in this category may be well studied, and its biology well known, but appropriate data on abundance and/or distribution are lacking. Listing of a taxon in this category indicates that more information is required and, acknowledges the possibility that future research will show that threatened classification for the taxon is appropriate (IUCN, 2001).

Not Evaluated (NE)

A taxon is Not Evaluated when it has not yet been evaluated against the criteria (IUCN, 2001).

All taxa listed as CR qualify for EN and VU, and all listed as EN qualify for VU. Together these categories are described as 'threatened', which form the heart of the system. So, for listing as CR, EN, or VU, there is a range of quantitative criteria; meeting any one of these criteria (and not whether all criteria are met) qualifies a taxon for listing at that level of threat. Because it is not clear at first which criteria are appropriate for a particular taxon, each taxon should be evaluated against all the criteria, and all criteria met at the highest threat category must be listed. A taxon may be transferred to a category of higher or lower threat if research indicates a change in their status (IUCN, 2001). The 'Documentation Requirements for Taxa Included on the IUCN Red List' are mentioned on the IUCN website.

Who Uses the IUCN Red List?

Information about species and ecosystems is essential for the sustainable use of our natural resources. The Red List is used by governments (wildlife departments and natural resource planners), the private sector, multilateral agencies responsible for natural resource use, and environmental treaties, for the latest information on biodiversity when making environment-related decisions. It is also useful for conservation-related NGOs, educational organizations, and others.

World Pheasant Association

The WPA has conducted valuable work for the conservation of galliform species, some of which are highly threatened. This work includes intensive habitat and population surveys, many of which have resulted in the establishment of national parks and protected areas, and increased awareness and education among people. WPA also works with private breeders and zoos around the world through Taxon Advisory Groups and other networks. The WPA-UK is a not-for-profit organisation and also publishes a journal called the *International Journal of Galliformes Conservation,* and online newsletters. Its Indian branch, WPA-India publishes a newsletter called *Mor* for its members. See 'Addresses' at the back of the book for membership information.

WPA is also the umbrella organization for the 5 merged Galliformes specialist groups – of cracids, grouse, megapodes, pheasants, and partridges, quails, and francolins - now united as the Galliform Specialist Group. This Specialist Group operates under the joint parentage of WPA and the SSC of IUCN. WPA will also compile the action plans for all the species of birds in this group, with information on the status, threats, and action required to protect the species, and make it available to conservation planners.

WPA was started largely by aviculturists, as were its chapters in India, Nepal, and Thailand. Only WPA-Pakistan, with its very strong affiliations to World Wide Fund for Nature (WWF), can be said to have started from conservation rather than avicultural roots (Howman and Sivelle, 1982).

Though some pheasant species have long been kept as ornamental aviary birds in Europe and USA, and at least one species (the ring-necked pheasant) has been bred on a large scale (for use in hunting sport), these activities have been recreational. The first attempt to orient such activities towards conservation was made when the Pheasant Trust was established in the UK in 1959, largely through the efforts of Mr. P. Wayre. The aviculturists, who had been keeping pheasant lines from the time they were first brought to Europe, were facing a problem of inbreeding in the pheasants, since it was no longer possible to get a steady fresh supply of pheasants from their indigenous range countries. Also, deforestation and poaching had caused the pheasant populations to decrease sharply in their natural habitats, and the captive pheasant population in the private aviaries of Europe and USA, offered a hope of reviving their populations by means of reintroduction into native habitats.

After a few years, some aviculturists in the UK wanted to widen the scope and the range of their activities, and hence the WPA was established in 1975, when a meeting of a few pheasant-enthusiasts was called by Keith Howman, an aviculturist in the UK. Very soon, the WPA established some overseas chapters and the one in India was formed in 1979, which had scientists from the Indian Veterinary Research Institute, Bareilly (Uttar Pradesh), and Zoological Survey of India as members. Since then, WPA-India has been trying to focus the attention of various government and non-government agencies, as well as the general public, on the need for conservation of pheasants.

The Indian Chapter of the WPA organized an International Symposium in Srinagar (Kashmir) in 1982, which reflected the concerns of pheasant conservation: biology, threats, and conservation action. At this point, the study of pheasants in the wild was still in its infancy. The Himachal Wildlife Project (in early 1980s) was important in conducting quantitative survey work and providing training for its many Indian participants. It yielded the first contemporary data in India on the status and ecology of western tragopan *Tragopan melanocephalus* and cheer pheasant *Catreus wallichii,* at that time believed to be among the rarest pheasants in the world. It also identified a site for the establishment of the Great Himalayan National Park in the state of Himachal Pradesh, representing the western Himalayan moist temperate forest ecosystem (Garson, 2007).

All this laid the foundation for the intensive studies of the ecology and behavior of cheer pheasants *Catreus wallichii*, white-crested kalij *Lophura leucomelanos* and other Galliformes, as doctoral training projects in Himachal, and in what is now Uttarakhand (1985-90). During this time, the first surveys were also undertaken in the northeast Himalayan states, where many of India's pheasant species are found (the state of Arunachal Pradesh has 11 species of pheasants, the maximum among Indian states), as well as in the northern state of Jammu and Kashmir. The founding of Wildlife Institute of India (WII) in Dehradun (1986), and the initiation of a Masters programme in Wildlife Science at Aligarh Muslim University (Uttar Pradesh, India) around the same time, provided the much needed institutional bases for an emerging cadre of experts focused on pheasant studies (Garson, 2007).

By 1990, it was possible to use the information gained about a few species to raise their profile as flagships for broader conservation action. The western tragopan *Tragopan melanocephalus* and the Himalayan monal *Lophophorus impejanus* have become the State Birds of Himachal Pradesh and Uttarakhand respectively. Blyth's tragopan *Tragopan blythii* was a symbol for forest conservation in Nagaland (Garson, 2007). See appendix 3.

A second WPA symposium was held at the WII in India in 2004, providing a platform for showcasing the voluminous and sophisticated work done. Studies of distribution have discovered a new all-white tailed subspecies of Sclater's monal *Lophophorus sclateri* in Arunachal Pradesh, and confirmed the presence of western tragopan in Uttarakhand. The impact of pesticides on Indian peafowl has been investigated, as has the extent of bushmeat hunting for pheasants in the western and northeast Himalayas. Conservation breeding of the western tragopan and cheer pheasants is being undertaken in Himacahal Pradesh (Garson, 2007).

The last WPA symposium was organized in November, 2010 in Thailand. The summary of the symposium was published in the new newsletter of WPA, *G@llinformed*, in which Dr. John Carroll (member of IUCN/WPA and wildlife professor at the University of Georgia, USA) has pointed out - "we are seeing a handing over of the academic and research torch from Europeans and Americans to nationals throughout the range of Galliformes". Apart from the papers from the symposium host countries (China in 1989 and Thailand in 2010), in the 1989 Symposium, there were 48 presentations and the primary authors on 33 of them were Europeans or North Americans; and in 2010, there were 43 presentations, and only 10 of them have the primary author from Europe or North America (Carroll, 2011). This reflects the growing numbers of scientists and government participation in conservation activities in galliform range countries. WPA scientists help to conduct research, monitoring, and management of pheasants in Asian countries.

Threats to the Survival of Pheasants

Why is the pheasants' habitat in danger? Why are their population numbers decreasing? What are the major types of threats faced by pheasant species?

Asia has a great diversity of habitats, ranging from Arctic tundra to tropical forests, and including the highest mountains in the world. This is reflected in the region's immense richness in birds and other wildlife. Asia also has a very large and rapidly growing human population, and dynamic economies. As a consequence, the region is experiencing environmental change, and many of its habitats and the biodiversity that they support are struggling to survive. Every country in Asia has populations of threatened birds, and needs to take action for the conservation of these species and their habitats (BirdLife International, 2001).

Forests are by far the most important of all the habitats for threatened bird species in the Asia region. The majority of threatened forest birds occur in the tropics (>90%) and in moist forest types (>80%), and the single most important forest type is tropical lowland moist forest with >70% of threatened forest species. Tropical montane moist, tropical dry, mangrove, and temperate forests also support some threatened forest birds (BirdLife International, 2001).

The Asia region has more than one hundred bird species which are now Critical or Endangered. Several of these have not been recorded in recent decades, and may have already disappeared. The issues that are driving these species towards extinction include mismanaged forestry and illegal logging, and conversion of forest land for agriculture and plantations, particularly in the tropical forests of Indonesia, Philippines, and mainland Southeast Asia. Urban development is affecting natural habitats, and includes major dam and irrigation projects, and the construction of roads into forest areas. Volunteer scientists of IUCN and BirdLife International are working to save the species, gathering data and making recommendations to governments –

▪   Identifying gaps in habitat-range protection of threatened birds, and proposing sites for the formation of new Protected Areas.

▪   Increasing funding, and proper equipment and training of forest guards and managers, are needed, especially for the sites for threatened biodiversity. Appropriate funding and training of the governmental departments responsible for conservation are also required.

▪   Conservation of many threatened Asian birds is hindered by incomplete data, and surveys or ecological studies are needed to address the gaps in knowledge.

▪   Proposals are made to reduce or eliminate the negative impacts of the major landuse issues affecting the threatened birds, and to control hunting and wild bird pet-trade (BirdLife International, 2001).

Habitat Loss and Degradation

For most of the 'threatened' pheasant species, the loss of their forest habitat is causing the decline in their populations. Tropical and temperate forests are being cut down to make way for agricultural uses, road building and urban encroachments, or for plantations such as timber, coffee, rubber, oil palm, and paper pulp, or are being degraded due to grazing of livestock, and fodder and firewood collection.

Since the early 1970s, the Southeast Asia - Pacific region has been the main source of the tropical timber trade – taking over from Africa, which supplied considerable quantities of logs to Western Europe during the 1950s and early 1960s. In the mid-1980s, 83% of tropical timber traded on the world market originated in the Southeast Asia- Pacific region. The scale of exploitation of forests in the last few decades has placed their future in peril. Timber is the major product. Forests are also cleared to grow plantation crops such as oil palm, coffee, and paper pulp (which make a major contribution to the region's economy), and for shifting cultivations of various kinds. Indonesia and Malaysia together produce 80% of the world's palm oil, which is used for food, cosmetics, detergents, toiletries, etc (Collins et al., 1991). It is exported throughout the world - European countries (16% of global palm oil use), India (23%), China (18%), and US multinational companies being the major consumers (while Indonesia and Malaysia together consume 32%) (WWF, 2013 b). [Oil palm *Elaeis guineensis* originated in Guinea, Africa, and was first introduced to Java by the Dutch in 1848. It was introduced to Malaysia (the then British colony of Malaya) in 1910 by Scotsman William Sime and English banker Henry Darby (Wikipedia)].

Fig 4.3 Photo of an oil palm tree with fruit.

Timber extraction by logging is the primary reason for deforestation. Logging is especially common in areas containing tropical forest on level ground, where commercially valuable trees can be easily extracted on a large scale. Much of the logged wood is used for making furniture, plywood, roofing (of Ironwood), and floorings for export to Europe, USA, and Australia, via Japan and China. The pheasant species likely to be affected are crested fireback *Lophura ignita*, crestless fireback *Lophura erythrophthalma*, and Bulwer's pheasant *Lophura bulweri* (in Sumatra and Borneo). Habitat loss and degradation due to various causes has also affected other pheasant species in Malaysia, Cambodia, Hainan Island

(China), China, Sri Lanka, and India (Fuller and Garson, 2000). [See blog site www.pheasantsoftheworld.blogspot.com for photos on this topic].

Fig. 4.4 A painting of a male Bulwer's pheasant from 'A Hand-book to the Game-birds' vol. I and II, by W.R. Ogilvie-Grant, 1897.

The area of natural forests within logging concession forests far exceeds that covered by protected areas. 'Logging concession licenses are granted by the government to logging companies to selectively choose and fell trees; and they have to ensure that the forests continue to grow and replace the trees that are harvested. Also, deforestation (or open clearing) is prohibited in logging concession area' (Wich et al., 2012). 'However, during the past 5 decades, the commercial extraction of roundwood (wood in the rough state) in logging concessions in Southeast Asia has proved to be particularly destructive. Forest management plans designed to ensure sustained yields have rarely been implemented. According to a study in 1990s, less than 1 % of natural forests in Southeast Asia undergo forest management. Numerous surveys had shown that in Southeast Asia, selective logging in logging concession forests destroys far more trees than just the ones that are extracted. Timber extraction even at the level of 10% of the trees destroys 65 % of the timber stand. A preoccupation with speed was much to blame. Logging was carried out with the maximum possible speed to ensure a profitable return on capital invested, particularly in heavy machinery. Very little forest road-maintenance was undertaken, and commercial species regeneration was rarely practiced. Bad felling practices and irresponsible management are widespread' (Collins et al., 1991). Conservation effort should focus on natural forest areas that have been selectively logged or are scheduled for logging. Logged forests are vulnerable to land-use conversion to oil palm plantations, or agricultural and urban uses, and this should be checked.

The Rio Earth Summit in 1992 highlighted the need for sustainable forest management to prevent rapid deforestation, particularly in the tropics. New tools to achieve Sustainable Forest Management have been promoted, such as guidelines and certification standards. SFM practices also include the development of national guidelines, training for forest managers, support from academic institutes and NGOs, and adequate law enforcement – that are followed to various extents in the countries of this region (Dennis et al., 2008). Two international organizations associated with the IUCN – 'The Rainforest Alliance' and 'Forest Stewardship Council' – have been working towards establishing policy and infrastructure for sustainable harvesting of rainforests in Africa, S. America, China, and Indonesia, since the 1990s. They have developed certification guidelines for products originating from rainforests and rainforest area plantations (of products like wood, tea, coffee, cocoa, and paper) for producers in these countries. They have also developed an advisory for forest management and improving supply chain mechanisms for products. The rights of the indigenous tribes affected by plantations are addressed; as is the matter of 'sustainable tourism' involving local people. There are also other organizations that help in education and awareness campaigns at the community level in these countries. Similarly, the International Tropical Timber Organisation (ITTO) has published a series of guidelines since the 1980s with the aim of maintaining the ecological balance in 18 tropical timber producer countries, and improving market conditions for the timber trade.

Philippines and Thailand have enforced logging bans, and Cambodia has also declared a logging freeze in 2002 to curb rampant deforestation. In 2010, Indonesia announced a 2-year moratorium on deforestation. During this time, Norway would contribute up to one billion dollars to fight deforestation in Indonesian forests by setting up a control mechanism, and offer an aid contingent for the capital city Jakarta's progress up to year 2014. [An estimate by the international NGO Greenpeace had placed the deforestation rate in Indonesia to be a forest area equivalent to 300 football fields felled every hour!]

Proper harvesting techniques (such as reduced-impact logging, logging restriction on slopes, along watercourses and adjacent to protected areas, etc), guidelines to conserve biodiversity in logging concession forests (habitats such as pools, wallows, saltlicks, and riverside habitats should be preserved, etc), landscape-level approach to land-use planning (including managed forests, protected areas, and other land uses), and a better understanding of forest economics are needed. There is much room for improvement in the sustainable forest management sector, and nations in Southeast Asia can achieve this through a network of public and private organizations and NGOs (including the IUCN, WWF and ITTO) (Dennis et al., 2008).

Apart from affecting the pheasant species, forest loss and fragmentation in Indonesia and Malaysia (1/3rd of which in the last 10 years was due to the expansion of oil palm plantations) is also affecting orangutans, pygmy elephants, and Sumatran tigers and rhinos. The Roundtable on Sustainable Palm Oil (RSPO) was established in April 2004 as a multistakeholder platform (including producers, processors, consumer goods makers, retailers, banks, NGOs, and WWF), to develop and implement global standards for sustainable palm oil. Its certification label (CSPO) is used by 14% of the global palm oil produced, and is growing (WWF, 2013 b). With growing awareness, market pressures have also created a demand for sustainably managed timber and oil palm plantations.

Hunting and Illegal Trade

Shooting and snaring of wildlife is common throughout much of Asia, by ethnic communities living in or near forests, whose lifestyle is based on shifting or semi-shifting cultivation and hunting, although pressure is less intense in Bhutan and Tibet for cultural reasons. The larger forest bird species are most seriously affected, notably the hornbills, partridges, and pheasants, which have been much reduced in many areas. In China, Thailand, Laos and Vietnam, larger-bodied birds have almost been hunted out, and snaring has reduced many forest-floor species to very low densities. The collection of eggs and chicks of birds for food is also frequent. The threatened species include several which are hunted because of their size, such as white-winged duck, crested argus *Rheinardia ocellata*, and green peafowl *Pavo muticus*. Green peafowl is often sold in markets for food, and live or dead birds are even traded internationally for their meat and feathers (in Vietnam, Laos and Java) (BirdLife International, 2003). In 2009, the status of green peafowl *Pavo muticus* in the Red List was changed from 'Vulnerable' to 'Endangered', as surveys revealed a very rapidly declining and severely fragmented population.

Galliform species have always been hunted for subsistence, and they are still hunted for food or trade. 'The impact of hunting is hard to quantify because much of it is illegal and covert. Nevertheless, hunting appears to have serious effects on populations of several pheasant species. All four of the pheasant species classified as Endangered appear greatly threatened by the activities of local hunters' (Fuller and Garson, 2000). It is estimated that 70% of all threatened bird species in Asia are currently taken for food, and 30% for the cage-bird trade. The keeping of birds for their beauty and song is popular in Asia such that many more species, even those with wide ranges, may qualify as globally threatened in the near future owing to the effects of bird trade.

Wildlife, such as the bearded pig, sambar, barking deer, mouse deer, primates, bats, crocodiles, pythons, megapode eggs, cave swift nests, grasshoppers and beetles have been a part of the diet of indigenous people in Southeast Asia. Birds such as hornbills, pheasants, birds of paradise, and bower-birds are hunted for personal decorations and ceremonies, as well as for the pot. Indonesia has a tradition of keeping cage-birds, and hunting is common in Philippines, and Nepal. Forest animals are the basis of a flourishing local and international trade, often to the detriment of the species involved. Live parrots, birds of paradise, crab-eating macaques, butterflies and freshwater fish etc are exported, many of them illegally.

Improved law enforcement is required to prevent illegal hunting both inside and outside protected areas. Control of gun ownership is meeting with some success in Laos and Vietnam, and could be expanded to other countries in the region, particularly near important protected areas. Education programmes concerned with forest conservation, threatened species and the hunting laws could help reduce hunting pressure, possibly by using the most charismatic threatened forest birds such as, Blyth's tragopan *Tragopan blythii*, or rufous-necked hornbill, as flagships. Community-based initiatives, including signing of stakeholder agreements with local households, and establishment of patrolling groups, have helped control hunting at key sites in Vietnam, and should be tried in other areas of high biodiversity value.

In the rainforest region that is home to the Congo peacock in central Africa, wars, ethnic conflicts, mining, and hunting have taken a toll on the pheasant's population. It is listed as 'Vulnerable' in the IUCN Red List. Democratic Republic of Congo (DRC) has 40% of the world's reserve of 'coltan', a metallic ore from which the mineral tantalum is extracted - an essential component of digital devices such as cellphones and computers. DRC also has large reserves of copper, cobalt, and diamonds, which are worth billions of dollars of revenue. Yet, the people of the country are extremely poor – a phenomenon that is known as 'resource curse' (a situation when countries that are rich in resources have poorer economic development than countries that have fewer resources). Also, since 1960, when the DRC was officially freed from Belgian colonial rule, wars with Uganda and Rwanda, and civil unrest have prevented infrastructural development in DRC. The country has adopted a new Constitution and a new National Flag as recently as 2005 and 2006. The country was being run by an interim government since 2003; and in the democratic elections held in 2011, people of DRC have chosen Joseph Kabila as their President; and though the country still witnesses violent conflicts, it is in a better condition due to the United Nations' participation. Ethnic conflict in Rwanda in 1994 had caused refugees to flee to the neighbouring DRC. The tropical forest in eastern DRC is now inhabited by the Rwandan refugees, increasing hunting pressure on the Congo peacock' (Birdlife International, 2014).

Insurgency was also a serious problem affecting wildlife in different parts of northeast India. In 1980s, insurgency had spread to many of the PAs such as Manas NP, Balpakram Tiger Reserve, Gumti WS and Sonai-Rupai WS. Poachers took advantage of the situation to carry out illegal hunting and logging.

The introduction of the Wildlife (Protection) Act, 1972 in India put an end to the legal hunting of Galliformes, yet they continue to be hunted, both for subsistence and sale across the country. Though trade in live birds is not large in quantity in India, trade in feathers (Himalayan monal crest, grey junglefowl hackles, blue peafowl feathers) has been reported frequently. Threatened species such as tragopans (western and Blyth's) are sold in markets or covertly; other threatened species such as cheer are hunted for local consumption (Kaul, 2007). Eleven of the Indian pheasant species are listed in Appendixes I, II and III of CITES. Thus, hunting of the species, as well as possession of skins, parts, or plumage, and local trade and international trade in live birds is prohibited. See Appendix 6 for details.

Fig 4.5 Fans made from blue peafowl train feathers for sale on a pavement in New Delhi, India. Photo by Anita Chauhan, 2010.

In 2013, the Ministry of Environment and Forests has decided to ban the trade in blue peafowl *Pavo cristatus* feathers in India. Thousands of peacocks are killed every year in the rural areas in India due to the use of synthetic pesticides, electrocution from electric supply-lines, and poaching for meat and feathers. The Wildlife (Protection) Act, 1972 had allowed the sale of peacock train feathers that have been naturally shed at the end of the breeding season, but since it is not always possible to ascertain whether the feathers are shed or have been plucked from poached birds, and because of the high demand which encourages poaching, the trade in feathers will now be banned. Many rural and tribal communities in India are known to consume peacock meat, and to sell the feathers for the lucrative trade. Peacock feathers are made into hand-held fans, earrings, and other decorative objects. Peacock feathers are also considered sacred and are used in religious functions.

India's wildlife is protected under the Wildlife (Protection) Act 1972 which applies to all Indian states except for the state of Jammu & Kashmir in northern India. According to the J&K wildlife laws, excluding the 18 bird species in the Schedule I of J&K wildlife law, hunting and trapping permits can be obtained for all other species from the Chief Wildlife Warden, after verification (Ahmed, 1997).

The Indian Wildlife (Protection) Act of 1972 [WPA 1972] is applicable only to Indian wild animals included in Schedule I to V. WPA 1972 bans the hunting and trade of all Indian wild birds, excluding those listed as 'vermin'. In 1990, export of live birds (indigenous birds and captive bred exotics) from India was totally banned. In 1991, local trade in birds was also banned (Ahmed, 1997).

Before the ban in 1991, the domestic and international trade in live birds was regulated by the Chief Wildlife Warden of the concerned states and the Ministry of Commerce. Now, import of birds is permitted by procuring a license for zoos, recognized scientific institutions, circus companies, and private individuals on the recommendation of the Chief Wildlife Wardens of the state government, subject to the provisions of the CITES. The CITES office in New Delhi coordinates the clearance of any bird brought to northern India (Ahmed, 1997).

Since 1991, there has been a complete ban on the trapping and trade of Indian wild birds, and yet illegal trade in wild birds is prevalent in all cities and towns of India. According to a report by TRAFFIC – India (the wildlife trade monitoring network – joint programme of WWF and IUCN) titled 'Live Bird Trade in Northern India', a survey conducted in 36 cities of northern India (in the states of Delhi, Punjab, Rajasthan, Haryana, and Uttar Pradesh) in 1997 revealed that atleast 250 species belonging to 51 taxonomic families were being traded in the market. The bulk of the trade was in parakeet species, munias, owlets, raptors, and wetland birds. Some galliform species, like various species of quails, grey, black and swamp francolins, red junglefowl *Gallus gallus*, kalij *Lophura leucomelanos*, and blue peafowl *Pavo cristatus*, are also trapped and sold. About 70 species of exotic birds that are been captive bred in India (eg. budgerigar, cockatiel, Java sparrow, and zebra finches) are also traded. WPA 1972 does not include the exotic birds and their trade in India is not regulated by any legislation (Ahmed, 1997).

The survey findings reveal that there are at least seven different purposes for which birds are captured. This perhaps explains why the bird trade is still so widely prevalent. Apart from the usual reasons such as pet trade, trapping for food, zoological laboratory uses, sport, and medicinal uses, birds are also captured for 'release trade' (ceremonial release of caged birds), and for black magic and sorcery (Ahmed, 1997).

In north India, birds are captured for the bird market by tribes which are almost exclusively dependent on birds for their livelihood. After the ban on bird trapping and sale, some of them have shifted to other professions such as tailoring and selling vegetables. However, a majority of them still continue the illegal trade (Ahmed, 1997).

A comprehensive picture of the bird trade in northeast Himalayas is yet to emerge. However, it is evident from preliminary surveys that the trade does exist, mainly for meat for local consumption. Even in some PAs such as Gumti Wildlife Sanctuary in Tripura and Pani-Dihing WS in Assam, the Galliformes are poached.

In a review of literature on hunting in north-east of India, published in the *International Journal of Galliformes Conservation* in 2011, sociologist Ambika Aiyadurai has advocated investigating the anthropological and socio-economic setting, along with the ecological approach, to balance hunting and conservation. Hunting among tribal populations in this region is not just an economic activity, or just a part of the local subsistence diet, but has a larger socio-cultural connection (Aiyadurai, 2011). The review describes the following trends.

Northeast India is rich in biodiversity containing the Eastern Himalaya 'biodiversity hotspot', and is also home to around 145 tribal communities, most practice shifting cultivation and hunting. Their dependence on forests for firewood, bamboo, medicinal plants, and other forest produce is high. Though data on local and indigenous hunting in India is very sparse, recent studies have gathered information about which species are hunted, and what quantity of meat is extracted. One study has reported that in the states of Nagaland, Mizoram, and Arunachal Pradesh, 134 species of animals (mammals, birds and reptiles) are hunted, including pheasants. Of the 50 Galliformes species native to India, 32 occur in the north-east of the country (Aiyadurai, 2011).

Although hunting in northern India had traditionally been the past-time of the wealthy, and the people's belief system treat animals such as elephants, tigers, and monkeys as sacred, the northeast region of India has a different belief system, and the population is largely non-vegetarian. Northeast region largely has tribal populations (for example, Nyishi community and Mishmi of Arunachal Pradesh) that live in close proximity to forests; and their relationship to forests shows a connection with guardian 'forest spirits' and a 'spiritual herdsman' who owns the forests. Deep forests are referred to as a 'spirit world'. Such beliefs are also followed by indigenous groups in other parts of Asia; for example, the Kerinci people of Sumatra, and tribes of Indonesia, and China. Hunting has been an important activity with a great pride attached to it. Gifting fresh or smoked wildmeat is a traditional norm and practiced during festivals. Wildmeat is offered as a bride price during weddings, and is regarded as a status symbol. Among the Apatani tribe in Arunachal Pradesh, religious rituals include offerings of wild boars, deer, and macaques. Nishi tribe use black bear skin and hornbill tail feathers to decorate their headgear, and animal skin for making sheaths for their traditional machetes and as shoulder belts. Mishmi people make hand fans from pheasant tails (Aiyadurai, 2011).

In the absence of employment options, a traditional activity like hunting becomes commercially beneficial for local people, especially, given the proximity to Myanmar and to wildlife trade routes in China.

Apart from human population pressure, it is the demand from urban populations for wildmeat, and the diffusion of new technologies of hunting, that are causing the change. A popular Hindu religious site 'Parshuram kund' in Lohit district (Arunachal Pradesh) is visited by a large number of pilgrims during January, for a religious fair. This fair has been used as a site for trading of local wildlife products, but lately the temple site has become the centre of exchange between outside traders (marwaris) and local villagers, in this case Mishmis. Local hunters look forward to this fair in January which also coincides with the hunting season, and provides a chance for hunters to sell their products at a profitable price (Aiyadurai, 2011).

The lack of economic alternatives in the region means the villagers remain dependent on wildlife hunting. Though the major consumers of wildmeat in Northeast are the rural communities, in Nagaland, high income families also eat wildmeat which they see as a luxury. Guns and locally prepared traps are used to hunt wildlife. Increased accessibility to sophisticated weapons and ammunition, and the growth of markets, has led to a shift from cultural value to economic value of wildlife products. In Nagaland, a Kalij pheasant *Lophura leucomelanos* is sold for Rs. 110 (£1.6) whereas in Walong (Arunachal Pradesh), pheasants are priced at Rs. 200 (£2.7). Earlier, a villager would share wildmeat with family members and villagers to maintain kinship ties, but with an economic value attached to it, wildmeat is often traded in the market (Aiyadurai, 2011).

An erosion of hunting taboos has also been seen. Hunters follow certain taboos which are related to conservation practice to prevent over-hunting of animals. Some species, such as the yellow-throated marten and hoolock gibbon are not hunted. Currently, there is a large scale conversion of Miju Mishmi to Christianity. The villagers who have converted have abandoned the ritualistic way of worshipping spirits, but they continue to hunt (Aiyadurai, 2011).

In remote areas of Northeast India, hunting still continues largely due to its linkages with local customs. The awareness of conservation and sustainability issues is extremely low. Most villagers perceive wildlife as an inexhaustible resource. Similarly in Vietnam, villagers do not seem to consider the extinction of certain species as a problem. This perception that 'wildlife is in plenty' is strongly rooted, and thus, communities may not participate actively in conservation projects. Thus, there is a need to orient the education programs to include facts about wildlife populations (Aiyadurai, 2011).

In order to regulate hunting in Arunachal Pradesh, erstwhile hunters are employed as field assistants in wildlife monitoring programmes; and villagers have taken a stand against hunting by taking a pledge. Conservation NGOs provide medical support, training in health care, and education to the community, and they encourage people's participation in wildlife research as part of 'community conservation'. In 2001, a wildlife NGO, Wildlife Trust of India, distributed fiberglass hornbill beaks to Nyishi tribal people, for whom the hornbill beak is a traditional symbol of valour; and WTI has also provided them the knowhow to make synthetic hornbill beaks. Many areas in Arunachal have seen such high hunting pressure, that the great hornbill has become extremely rare. The Village Development Councils also announced a fine of Rs. 5,000 (£69) for any person caught hunting hornbills. In 2012, WTI has also provided to the tribal people, accessories made from synthetic fur, as an alternative to bear fur, during the annual Hornbill Festival in Nagaland (Aiyadurai, 2011). Also, encouraging tourism activities, and commercial quail farming, will provide an alternate livelihood option to many people in the region.

Also, infrastructural and economic development in Northeast India is an important issue. There are about 100 hydroelectric dams planned in Arunachal Pradesh, which could adversely affect the state's biodiversity.

Human disturbance

Harvesting activities other than hunting also affect pheasant populations because of their predominantly ground-feeding and –nesting habits (Fuller and Garson, 2000). The collection of gucchi (a.k.a. morel mushroom, *Morchella*) in the western Himalaya in India coincides with the spring breeding season of western tragopan *Tragopan melanocephalus*, and often may cause hens to abandon nests, or result in eggs being stolen. Guchhi is one of the Minor Forest Produce that can be collected from forests outside the core zones in PAs, after obtaining permits from the Divisional Forest Officer. At the fairs where guchhi collectors sell the mushroom (eg. *Banjar* fair in Kullu district, and *Lavi* fair in Shimla district, Himachal Pradesh, India) along with other medicinal herbs and roots from the forests, the selling price is Rs. 7000/kg, and this price increases to Rs. 20,000/kg when it is sold by dealers in Delhi and Mumbai. The price is even higher in the export market. The collection and trade of this mushroom, which only grows in the wild, is unregulated. The Forest Corporation in the state of Himachal Pradesh is in the process of setting up a trading center at Shamshi, and formulating rules to regulate the trade (Chauhan, 2011).

Similarly, medicinal plants collectors and ringal bamboo collectors are potential sources of disturbance. Many of the medicinal plant species that grow in the moist temperate forests of the Greater Himalayan zone in western Himalayas (many of them terrestrial orchid species) are also listed in the CITES so that their export is regulated, and yet, over-exploitation of these plants has caused populations of several of the species to decrease. Also, during spring season, shepherds migrate with their flocks of goats and sheep to the alpine pastures through these forests. Other species of pheasants are also affected by anthropogenic disturbance.

Fig 4.6 A Gaddi shepherd taking his herd of goats and sheep to alpine pastures, along the NH 22 in Himachal Pradesh. Photo by Anita Chauhan, April 2011.

Fig 4.7 A. Medicinal herb *Viola canescens* and fresh morel mushrooms, in Kotgarh, Shimla district, Himachal Pradesh, B. Dried morel mushroom for sale in Lower Bazar, Shimla Mall, Himachal Pradesh, India. Photos by Anita Chauhan, April 2011.

An international standard for the sustainable harvesting of Medicinal and Aromatic Plants was developed by the IUCN Medicinal Plant Specialist Group and other organizations [TRAFFIC, WWF, and BfN (The German Federal Agency for Nature Conservation], published as the ISSC-MAP Ver 1.0 in 2007. This was revised and combined with the FairWild Standard for natural products in year 2010 (which is developed by the FairWild Foundation, a non-profit Foundation based in Switzerland), and is now followed by a few countries, including India, as FairWild Standard Version 2.0. It is used as the basis of a certification and labeling scheme, as well as to provide a framework for resource management in other scenarios. In the FairWild Standard, the 'Principles and Criteria for Collection Operations' state that negative environmental impacts caused by collection activities on other wild species, the collection area and neighbouring areas shall be prevented (FairWild Foundation, 2010). There is an urgent need to rigorously implement this standard, so that plant species and habitats can be saved from over-exploitation in the western Himalayas. Plant extracts are also used in the making of cosmetics, perfumes, soaps, beverages, etc. The concerned government bodies in the central Ministry of Health and Family Welfare, India are – the National Medicinal Plants Board and the Centre for Research, Planning and Action. See 'Further Reading' at the back.

Fig 4.8 The FairWild label.

Invasive Species

Invasive species affect 10% of all threatened bird species in Asia, the majority inhabiting small islands. For example, feral dogs, cats, and rats are speculated predators on Indonesian islands, while Javan mongooses and Siberian weasels are additional pressures to the threatened species in Japan. Introduced diseases via escaped cage-birds to wild populations are potential threats to some species.

Pesticides

Agricultural development and changes in land-use have caused a widespread decrease in bird species through habitat loss, modification, and fragmentation, particularly through the conversion of large areas to intensive crop cultivation, irrigation schemes, increased pesticide usage, and livestock-grazing. Farmland birds such as the Indian peafowl (as well as Great Indian bustard, francolin and quail species), are adversely affected. The intensification of agriculture in India has caused increased use of organo-chlorides and organo-phosphates. An alarming number of blue peafowl *Pavo cristatus* mortalities have been attributed to pesticides across India (Andhra Pradesh, Punjab, Karnataka,

Tamilnadu, Maharashtra, Haryana, Gujarat, Uttar Pradesh, Orissa, etc). It is frequently reported in the newspapers. Peacock deaths are either caused by grain mixed with pesticide that is sown, or the spraying of poisons on to cotton crop causing peacock death (over 70% of the amount of pesticide used in India is applied to the cotton crop), or due to electrocution from new power lines. An estimate states that at least five birds die of electrocution and pesticide poisoning each day in just two Haryana districts. Only seven percent of peacocks die of natural causes in India (Chauhan, 2013).

In a study conducted in Keoladeo Ghana National Park, Rajasthan, India in 2011, pesticide residues and metabolites were found in the soil from the agricultural fields in the vicinity of the park, as well as soils inside the park which receives run-off from the fields. Therefore, legislation for a buffer zone in the vicinity of protected areas, which uses minimum synthetic pesticides, may be required (Bhadouria et al., 2012).

The All India Network Project on Agricultural Ornithology has developed certain bird scaring devices, and has recommended them to farmers for protecting crops from bird damage. Reflective ribbons can be used to scare away various species of birds from crops like sunflower, maize, guava, grapes, paddy, ground nut, onion, lady finger, pomegranate, dates, apple, kinnow, sorghum, pearl millet etc, and from fish ponds. However, this deterring device is not yet being extensively used in India. The reflective ribbon is made of polypropylene, and is coloured red one side and silvery-white on other side. These colours make the ribbon highly reflective and it shines brightly in the sunlight. Lengths of the ribbon are fixed with the help of poles at a height of about one foot above the crop level. During daylight hours, bright reflections from the ribbons and humming noise produced as the ribbons flutter about in the wind scares the birds from the field. The cost of one 100 meter long roll of reflective ribbon is only Rs. 45 (US $ 1). Bioacoustics (broadcasting of recorded distress calls of parakeets and other depredatory birds), and botanical repellant sprays (derived from neem and tobacco leaves) on crops, are also used to deter birds from crop fields (AINPAO, 2013). Such methods, combined with other Integrated Pest Management methods (such as trap cropping, and lure cropping), result in economic benefits to the farmers, and provide a sustainable output without disrupting the ecosystem. Presently, only 2% of agricultural land in India is under the IPM system. IPM technologies have been developed by the National Center for Integrated Pest Management of ICAR in the Ministry of Agriculture. NCIPM also maintains a crop pest database, and conducts pest and disease surveillance and advisory projects (Chauhan, 2013).

In Europe, there has been a significant decrease in the (introduced) wild ring-necked pheasant populations in the last 40 years. This has been caused by agricultural intensification and widespread use of insecticides and herbicides, causing habitat loss and reduction in food sources on farmland. Previous studies in the 1980s, on European farmland wild gamebirds like ring-necked pheasant *Phasianus colchicus*, red-legged partridge, and grey partridge, have shown that home range size is smaller if the habitat has weedy areas, grasslands and abundant insects. And the number of chicks surviving in such habitats is more. A new study conducted in a large commercial farm in Austria has found that, planting of rotational fallow strips of land (or set-aside land) on the edges of fields (but away from woodlands as they harbor predators) with 'brood-rearing seed mixtures' containing seeds of plants such as native wildflowers, sunflowers, mustard, lucerne, cereals, and grasses, provides an appropriate cover and food source for rearing of pheasant broods. No herbicide or pesticide is used in these set-aside areas. Radio-telemetry (years 2001 to 2003) of pheasant hens with broods has shown that broods that incorporated set-aside areas in their home-range had 100% survival. Also, recommendations have been made regarding the density of plants in set-aside areas, and providing strips of open land in the set-aside area to maximize suitability of the habitat (Draycott et al., 2009). Thus, it may be recommended that farmers should incorporate rotational set-asides in their farms; and appropriate seed mixtures that can be planted in these set-asides should be made commercially available, in order to provide insect-rich foraging areas to farmland birds. (Also see Appendix 5).

In Situ Conservation

In India, the wildlife is protected through a network of Protected Areas - natural habitats consisting of national parks, wildlife sanctuaries, sacred groves, community forests etc. The first national park in India was declared in 1935, now famous as the Jim Corbett National Park. Since 1947, there has been a steady rise in the number of Protected Areas especially after the enactment of the Wildlife Protection Act in 1972.

Table 4.2 Some statistics about the Protected Areas in India.

Year	Number of National Parks and Sanctuaries	Area in sq. kms.	% of India's geographical area
1988	54 NP, 372 Sanctuaries	109,652	-
2000	566 NP & Sanctuaries	153,000	4.66
2010	597 NP & Sanctuaries	154,572	4.74
Recommended by WII	870 NP & Sanctuaries	1,88,764	5.74

Recently, the Bombay Natural History Society, in collaboration with the government and various NGOs, has identified 463 important bird areas (IBAs) in India. Out of these 463 IBAs, 199 are not officially protected. Many of these IBAs are very important for bird and other biodiversity protection, and should be included in the PA network. Similarly, the Wildlife Trust of India (an NGO based near New Delhi, India), along with the Asian Elephant Research and Conservation Centre, has identified 88 elephant corridors outside the PA network that also need protection.

Besides the official PAs, there are numerous sacred groves scattered all over the country, which are important for biodiversity conservation. Some sacred groves represent forest types that have disappeared from the area. Besides sacred groves, there are many small community conserved areas. Many villages do not allow hunting in their village ponds and lakes. These serve as excellent habitats for waterfowl. Similarly, the tribal reserves of Andaman and Nicobar Islands are perhaps the best-protected forests on those islands.

The present Protected Area network has many inadequacies. Several biological regions, communities and species are not or only partially represented, and most of the PAs are too small in size to give long-term viability. This could lead to genetic isolation of small populations and result in populations becoming unviable, endangered by all the classic threats of an island biogeographic situation. There is thus an urgent need that the sanctity of the Protected Areas, along with their surroundings and linkages, are preserved.

In rainforests, which are characteristically species rich, each species tends to have relatively few individuals in any given part of a forest. Therefore, in order to maintain sufficient individuals to comprise what biologists consider a viable breeding population, substantial areas of protected forests may be required. One study pointed out that areas in excess of 2000 sq. km need to be preserved to maintain populations of large frugivorous animals such as gibbons, and hornbills, which occur at densities of less than 5 individuals per square kilometer (Collins et al., 1991).

In the Himalayan states in India, there is a strong need to focus university level research and training towards pheasant conservation. There is a pressing need to upgrade the infrastructure in Protected Areas, including equipping the forests guards, and providing adequate means of communication and guard huts to improve patrolling and prevent poaching. Forest management practices should focus on making the habitat more conducive to birds, especially the galliform species. Man-animal conflicts in the western Himalayan states, especially those involving Himalayan black bear and leopard, need to be addressed through research, and rescue and rehabilitation of the animals. 'One such project has been initiated by the Wildlife Wing of the Himachal Pradesh Forest Department in collaboration with experts from the Wildlife Conservation Society of India, Norwegian Institute for Nature Research, and the Large Carnivore Conservation Laboratory, Washington State University. The 5 year study (2013-2018) will use camera traps, radio-collars, DNA and diet analysis, etc., to map and assess the conflict situation with leopards in the state, and suggest solutions to the problem' (Chauhan, 2013).

How is a National Park Managed?

The protection of wildlife habitat is the most important conservation tool for protecting the wildlife. The Forest and Wildlife Departments of each State in India ensure this through adaptive ecosystem management practices.

The management of a national park ecosystem involves – (a) building a resource inventory, (b) monitoring of specific components, (c) management action, and (d) research. It will be interesting to find out how the various components of a national park are surveyed, selected as indicators for monitoring the quality of the habitat, and the problems that are encountered, solved through research. In this way, a healthy ecosystem can be maintained for perpetuity, for the enjoyment of our future generations (Jenkins et al., 2003).

(a) A resource inventory, as the name suggests, is a quantitative record of the presence, abundance, and distribution of resources (geology and soil, biogeochemical cycles, climate, vegetation communities and fauna) in the national park. Thus, inventories document the species occurring in the park and their distribution. The historical data pertaining to human use of resources in the park, anthropogenic threats, and natural disturbances are also documented (Jenkins et al., 2003).

(b) Monitoring is very important for the scientific management of national parks and sanctuaries. Ecological monitoring is the sequential measurement of ecological systems over time with the primary purpose of detecting trends in the park ecosystem (Jenkins et al., 2003). The basic monitoring aspects of an ecosystem-based monitoring program include:

1.   Ecosystem drivers (eg climate and human pressures)
2.   Indicators of ecosystem health (eg biogeochemical indicators)
3.   Known threats (eg impacts of exotic species)
4.   'Key' species (Jenkins et al., 2003)(eg rare, or threatened species such as pheasants, or snow leopards)

These aspects provide a baseline to detect an 'unnatural' change in park resources, and provide the earliest possible warning of unacceptable change (Jenkins et al., 2003).

 Thus, a monitoring program helps in establishing a reference condition or state of the natural resources, from which future changes can be detected. It helps park managers to identify problems, formulate management plans, and assess their effectiveness (Holling, 1978). Issue-specific monitoring programs are also important because they provide the basis for evaluating the effectiveness of specific management actions.

Monitoring and resource management (a.k.a. 'conservation management intervention') help to conserve and achieve a more sustainable use of the biological resources. Interestingly, a recent study published in the *Journal of Applied Ecology* has shown, that involving local people from the communities dependent on forest resources, in data collection and analysis for environmental monitoring, improves the speed of solving resource management issues. The study examined 104 published environmental monitoring schemes, and found that scientist-conducted monitoring typically takes 3-9 years to implement, while monitoring schemes that involve local villagers (a.k.a. participatory monitoring) take 0-1 years to implement (Danielsen et al., 2010).

In another study, 2 types of monitoring methods (used by local people and scientists) were compared across 1.1 million ha of protected areas in the Philippines. The 1st type – included a 'focus group discussion method' - that involved the establishment of volunteer community monitoring groups (made up of 5–8 local residents) who collected information on resource use, species populations, and habitat condition on a regular basis, and met the protected-area staff 4 times a year for discussions. The members of the community monitoring groups were identified in cooperation with local community leaders so as to include the most experienced hunters, forest product gatherers, and fisher-people in each village. This method was intended to indicate changes in the perceived harvest volume per unit effort. The 2nd type of monitoring method was the standardized techniques used by scientists - the 'fixed point photography' and the 'line-transect method'. These were carried out by the protected-area staff without the involvement of local communities. Scientists found that the motivations for community members to participate in monitoring work are primarily linked to the socioeconomic benefits of monitoring, which range from enhanced *de facto* user rights to land and resources, to local status, pride in an area, and potentially enhanced training opportunities. That is, participatory methods are only continued when local people benefit from participation in the monitoring. Monitoring schemes that simply offload the costs of monitoring onto the local communities without assigning benefits are unlikely to be sustained. Despite the limitations, the study suggests that, from a government perspective, participatory biodiversity monitoring is more effective than conventional monitoring in terms of generating local management interventions in Philippine protected areas. The interventions are relevant for biodiversity conservation because they address the most serious threats to the biodiversity of the area, and led to changes in local policies with potentially long-term impacts. Also, although participatory monitoring involves an initial cost of training and capacity-building, and the accuracy of participatory monitoring and decision-making may not always be good, the study findings suggest that promoting 'community-based' and 'citizen science' approaches would link environmental monitoring to awareness raising and improved decision-making in resource management. These approaches could also speed-up the progress toward achieving the goals of international agreements, such as the Convention on Biological Diversity. A few agreements, such as the Ramsar Convention on Wetlands of International Importance, are already using thousands of volunteers around the world to collect required data on bird populations, which helps to compute the status and trends of wetland-bird populations (Danielsen et al., 2007, and 2013).

Let us look at another example concerning protected area conservation. The Great Himalayan National Park in Kullu district of Himachal Pradesh, India has a total of 832 plant species belonging to 427 genera. The park is situated within one of the globally important "Endemic Bird Areas", and has 183 bird species including 5 pheasant species. Owing to the wide variety of habitats, rarity and endemism of biological diversity, and geological beauty, the GHNP has been nominated as a 'World Heritage Site' by UNESCO. From the time of its establishment in the 1980s, the Great Himalayan NP has been the focus of attention for the Himachal Pradesh Forest Department. The Park administration has implemented several schemes to reduce the dependence of people who live in the areas fringing the Park on the park resources. An eco-development zone was set aside at one of the boundaries of the Park so that the villagers could utilize the forest resources in a planned way, without disturbing the core area of the Park. Thousands of families inhabit the villages and hamlets present in the eco-development zone. The villagers have depended on the forests for the following resources –

(1) Fodder, and grazing of herds of sheep and goats, (2) Edible and medicinal plants collection for local and commercial use, (3) Morel mushroom collection for commercial use, (4) Bamboo for mat- and basket-weaving for local and commercial use, and (5) Fuel wood, and timber for the construction of houses (Vasan, 1998).

The eco-development zone is able to supply the villagers' needs of fuel wood, fodder and timber. The Park administration has provided help to the villagers to grow medicinal plants and other cash crops, so that they do not collect these from the wild. Collection of the highly prized medicinal plants and morel mushroom still remains attractive for the poorer households (Pisharoti, 2008). The Park administration has started several schemes, such as - Biodiversity Conservation Society (BiodCS for park management), Eco-Development, Eco-Tourism, Women's Small Credit Groups, Alternate Income Generation, etc. – and their persistent efforts have effectively reduced the villagers' dependence on the forest. Several NGOs are active in the area, such as 'Biodivesity Tourism and Community Advancement' (BTCA), 'Friends of GHNP', and 'My Himachal', that provide help to the villagers in various forms.

The Wildlife Wing of the Forest Department of the state of Himachal Pradesh (India) has recently embarked upon long-term ecological monitoring (LTEM) studies in the Great Himalayan National Park. Sampling plots, to gather LTEM information on species and aspects of habitat, have been laid on fixed permanent transects. Understanding the park issues and key attributes of the park ecosystem, identifying the monitoring indicators and sampling requirements, will help shape the monitoring program. The LTEM database collected will be used as a benchmark, and to highlight the diversity of the park.

Additionally, in and around many protected areas, villagers maintain traditional community-based systems for controlling and monitoring access to resources. Community leaders and people regularly discuss the availability and quality of natural resources. The participatory monitoring process benefits from these existing informal monitoring systems. For example, in the Panchayat System of governance in Indian villages, an elected group headed by a 'Sarpanch' controls the use of Community Forest resources; and in the state of Himachal Pradesh, India, the 'Sarpanch' now has the same powers as a Divisional Forest Officer (DFO) of the Forest Department, in deciding Community Forest resource-use by the villagers. However, *de facto* user rights may not always ensure the local community the ability to protect and use forest resources sustainably. In Kullu and Mandi districts of the state of Himachal Pradesh, India, the villagers have traditional rights to exploit forest trees for building/repairing houses and temples; and these rights are still in place (except in the protected areas), thanks to the Anderson Settlement Report of 1886 (Vasan, 1998) and the Wright's Settlement Report of 1917 (Hobley, 1992) respectively, which are prevailing. Each village household can exploit a certain number of live standing trees (usually deodar *Cedrus deodara* and kail *Pinus excelsa* are used for house- building) after seeking permission from the designated officials (Vasan, 1998). A traditional stone-and-wood village house requires about 10 trees for construction (Vasan, 2000), and a large percentage of the districts' population is rural. The high altitude forests in these districts are over-exploited, with the effects spilling over into the protected areas. How much of this exploitation is due to overpopulation, a poor Panchayat System, bad resource management, and the very active timber mafia, remains to be quantified. Even so, in the year 2009, the Himachal Pradesh Forest Department has raised a 'green task force' (Forest Police and forest police stations or 'van thanas') to prevent illegal felling and smuggling of timber worth several lakh rupees every year. The districts of Shimla, Sirmaur, Chamba, Mandi and Kullu in the state have reported the maximum number of cases of timber smuggling. The wood furniture manufacturing units in neighbouring Punjab and Haryana seem to drive most of the illegal trade. Himachal Pradesh is the 2nd state in the country after Madhya Pradesh, where the Forest Police has been introduced to check deforestation. The state of Jammu and Kashmir has also raised a Forest Protection Force recently, to combat timber smugglers and encroachers. A certification scheme for wood furniture and tax benefits for complying units can be introduced as a regulatory mechanism. Apart from this, deployment of 'wildlife watchers' from the local community to combat poaching, and regulation of herds of sheep and goats by issuing of permit licenses to the graziers, limiting the size of herds, vaccinating the cattle, and rotational closure of pastures, are measures that are needed.

The GHNP administration, headed by Mr. Ajay Srivastava, CCF, organizes capacity building programmes for the women and youth of the eco-development zone, to teach them aspects of eco-tourism and production of handicrafts. Production and marketing of wildlife-themed handicraft souvenirs that can be sold to tourists at the Park and at the State emporia, and setting up of community colleges that offer certificate level courses in tourism, ecology, photography, painting, foreign languages, etc., would also be useful in employing the village youth in the tourism sector. With these measures, the Great Himalayan National Park is moving towards its goals of conservation while addressing the traditional rights of village communities so that sustainable solutions may be found (Chauhan, 2013).

(c) Management action in protected areas involves the protection of the flora and fauna against anthropogenic pressures such as, poaching of wildlife, illicit felling of trees, forest fires, and illegal livestock-grazing. PA management also encompasses habitat management (non-consumptive management of crucial habitat units such as food, water and cover, to maintain diversity of wildlife and ecological processes), tourism management, relocation of villages situated at critical locations, ecodevelopment, and wildlife rescue. Development and maintenance of the administration, communication, and protection infrastructure is another important management activity in protected areas.

An example is the use of mobile devices (GPRS/Wireless broadband technology) in data gathering and information dissemination. The Forest Department of the state of Madhya Pradesh (India) has developed a mobile IT system called 'M-Mantra for Forest and Wildlife Management', enabling effective monitoring of forests and wildlife. The information originating in mobiles and computers is synchronized with a server through the communication network. Several applications for personal digital assistants (PDAs a.k.a. palmtop computers) have been developed, such as – a Fire Alert Messaging System (FAMS), a Wild Life Management System, and systems for Forest Navigation, Forest Planning and Geo-mapping. These are being used by thousands of field officials in M.P., for activities such as capturing wildlife sighting data with geo-coordinates through PDAs. This data can be used for better wildlife habitat management. The system also facilitates monitoring the activities of patrol parties (GoI, MP Forest Department, 2012).

(d) Ecological research involves measuring ecological components to explain causes and effects of spatial or temporal changes in ecosystem health, and also helps to explain complex relationships in ecological systems. In general, monitoring is the tool used to identify whether or not a change has occurred, and research is the tool to determine what caused the change (Jenkins et al., 2003).

For example, to develop an effective conservation strategy for protecting a gamebird species, it is often essential to know its detailed habitat requirements. A general idea of the habitat type that a species occupies (forest/ agricultural land/ marshes etc) is not sufficient. Scientists need to know how it uses habitat types within these areas, and which of these are most important for its survival. Radio telemetry is used for this, to gain details of the bird's locations. By gathering repeated radio-locations of a tagged bird, radio-telemetry enables the scientists to find out which habitat type is used most, and for what type of activity (Dowell et al., 1992).

In another example, the continued survival of the endangered snow leopard (*Uncia uncia*) is threatened by poaching for its pelt, retaliatory persecution by herders whose livestock (sheep, goats, yaks, horses) it preys upon, and the depletion of its natural prey (ibex and blue sheep) populations. Pastoralism is the dominant landuse over much of the snow leopard's range, and competition between wild ungulates and livestock is a major conservation concern for the species. In an attempt to find out the reasons and extent of man-animal conflict in the Spiti region in Indian trans-Himalayas, Himachal Pradesh, scientists (from the National Conservation Foundation, Mysore, India) have found that there are several factors that are bringing the snow leopards and wolves in conflict with the human agro-pastoral communities. Among them are – the socio-economic changes causing increase in human and livestock population, and unplanned development of tourism industry causing pollution and habitat degradation. Also, in a comparative study of 2 agro-pastoral communities in the Spiti region, it was found that the level of economic dependence of a community on the livestock, determines the *intensity* of the conflict (how many snow leopards are killed), though the actual *extent* of the depredation (how many livestock are killed by snow leopards and wolves) depends on the relative densities of livestock (sheep, goats, yaks, horses, etc) and wild prey (ibex and blue sheep) populations (Mishra, 1997).

A new study by the Snow Leopard Trust and National Conservation Foundation, Mysore researchers in the Upper Spiti landscape of northern India, has found that in areas with a higher number of wild prey species, there are actually *more* (not less!) livestock lost to snow leopards than in areas where natural prey is scarce. An increase in wild prey, it appears, supports a growing number of snow leopards; and while they may prefer ibex and blue sheep, the snow leopards will still hunt domesticated yak and horses occasionally, which means more livestock loss for the local herders. So, while a recovery of endangered wild prey species is undoubtedly a good thing, both for the snow leopards and for the entire ecosystem they live in, it was wrong to assume that this recovery would also solve conflict between snow leopards and herders. Instead, it might escalate those conflicts, as more snow leopards will be able to survive on a steady but varied diet of wild prey and livestock (Snow Leopard Trust, 2013).

Therefore, we see that each new scientific study reveals new information that will help to further improve the conservation programs. Studies on estimating the population of snow leopards and wild prey and livestock have been conducted, and finding out the carrying capacity of the ecosystem for the wild prey and livestock in Spiti appears to be the next step, after which the grazing niche could be divided between a suitably large wild prey population and the livestock. This way, a limit can be set on the livestock numbers. Additionally, insurance schemes and increased protection to livestock, and alternative employment to reduce dependence on livestock will help to keep the herders happy (Snow Leopard Trust, 2013).

Monitoring of Pheasant Populations in India

Galliforms are an important component of the wildlife diversity in India, and are well represented by pheasants, partridges, quails, francolins, snowcocks, and megapodes. They occur in a variety of habitats ranging from the hot and arid western India, to the cold and wet high altitude forests in the western Himalayas and northeast Himalayas; from the cold deserts in the north to the coastal plains in peninsular India. Galliform species are good indicators of habitat quality as they depend substantially on ground layer vegetation for food and cover requirements.   They form a significant prey base for a variety of predators – mammals, reptiles, and raptors (Sathyakumar et al., 2007).

As explained earlier, many of the Galliformes are threatened due to poaching for meat or feathers, and habitat loss due to the expanding human population. The long-term conservation of galliforms, and of pheasants in particular, is of high priority for many of the Indian states. In order to conserve and manage any wildlife species, basic information on its distribution, population, habitat use, and behavior is required (Sathyakumar et al., 2007).

Estimating the distribution and abundance, and monitoring of galliforms in India has been a difficult task due to the following reasons: (i) some of the galliforms inhabit rugged and remote high altitude regions or dense forests with thick undergrowth, (ii) most of the Galliformes are shy and cannot be observed easily, and (iii) some of them occur in very low population densities in nature. Nevertheless, monitoring of galliform populations is crucial. To begin with, information on presence/absence needs to be collected. After establishing the baseline information, monitoring should be carried out by recording the abundance and density on a regular basis (Sathyakumar et al., 2007). The following simple techniques for monitoring are proposed for the field staff of forest and wildlife departments, as well as for amateur bird watchers –

1.   Presence/Absence   Mapping: In   order   to   record   the presence/absence of a species for a sampling unit, the forest field staff can either use a compartment based map, or divide the area into grids or small units based on natural land features. For areas that are outside the PA network, one can record presence/absence of species at village, panchayat, block, and district level. Information can also be recorded from institutional campuses and other private land. Details such as the name and location of the sites where the species are being recorded should be maintained. The GPS location, altitude range, and general forest or habitat types should also be recorded for the site.

Confirmation of the presence of galliform species could be based on direct sighting, or evidence such as feathers sighted or calls heard, and reliable   secondary   information   based   on   published   information, department records and interviews with local people. Each species in a sampling unit is assessed qualitatively by defining categories such as absent, very rare, rare, common, fairly common, abundant etc., based on the number of individuals seen per day of field work (Sathyakumar et al., 2007).

Fig 4.9 A blue peafowl feather on the ground at the Yamuna Biodiversity Park, New Delhi. Photo by Anita Chauhan, December 2010.

2.   Encounter Rates: Encounter Rate (ER) is a simple index for abundance estimation and is expressed as 'number of individuals of a species seen per unit effort'. The unit effort could be time spent in intensively searching for animals in an area or it could be the distance traveled in an area intensively searching for animals. Number seen could be based on direct evidence (sighting) or indirect evidences such as calls, droppings and other signs such as digging.

One could survey an area for galliforms by walking along existing roads, trails or along a predetermined transect using a compass or GPS. Driving along roads is another way for surveying galliform species in an area. If the distance traveled is measured, then ER is calculated as ER = number seen/ km walked. In case the distance traveled is not known, one can use the time spent in searching the area to calculate ER as ER = number seen/time spent. For example, if a person spent 2 hours in a forest area intensively searching for pheasants and sighted 1 Tragopan, then ER = 0.5 tragopan/hour search. Encounter Rates are useful for monitoring the abundance of galliforms in an area, if done regularly (monthly/seasonally/annually). Adequate number of walks per month or season is necessary for calculating mean ER and standard errors. This technique is applicable for most of the galliforms (Sathyakumar et al., 2007).

For example, an exercise was conducted by the Park Managers and field staff of the Great Himalayan National Park (Himachal Pradesh, India) in May 2010, to evaluate the population abundance of the western tragopan *Tragopan melanocephalus*. In the 3 ranges in the GHNP – Thirthan, Jiwanala and Sainj – 15 Call Stations (based on records of previous sightings, their vegetation characteristics etc) were demarcated, each with a diameter of 600m and altitude range of 2528-2916 meters. Two field staff in each Call Station were given data sheets and a GPS to record the GPS coordinates, altitude and calls till 1 hour after sunrise. Each call was counted as a breeding pair. After analyzing the data collected, the encounter rate in GHNP was found to be 1.87 birds per call station, and the density was found to be 6.6 birds per square kilometers in the sampled habitat (Srivastav, 2010).

3.   Line Transects: Line Transect is a simple method used to obtain density estimate for galliforms in an area. In this method, one walks along a straight line and counts birds on both sides of the line. Line transects could be permanently marked for regular sampling. At least 2 or 3 transects of length ranging between 1-3 km should be laid in each area, and walked at least 2-3 times in a month during the early morning hours. This technique is best suited for pheasants such as monal, kalij, red junglefowl, grey junglefowl, and Indian peafowl. Statistical softwares (for example, 'Distance') are used to analyse the data for calculating ER and density estimates (Sathyakumar et al., 2007).

4.   Call Counts: The abundance of some pheasant species during breeding season can be estimated by using Call Count Technique. During April-May, males call during early morning hours to attract females and also to challenge rival males in the vicinity. Observers record the calls along a line transect in the pheasant habitat. This is an index of the number of calling males in an area. One can playback a recorded call in an area to get response from individuals in that area. Additional information on the group sizes during the breeding season will help in understanding the population size in an area. This method is best suited for tragopans, koklass and cheer pheasants (Sathyakumar et al., 2007).

Pheasants are found in various national parks and sanctuaries, as well as in areas outside the Protected Areas network. Management of pheasant habitats involves reducing human activities such as hunting, livestock grazing, and development activities such as plantations, mining, road construction, dams, etc. Stakeholders range from tribes, villagers, businesses, and industrial houses, but workable solutions have to be found to save the species and their habitat (Garson, 2007). Over the past 25 years, a substantial amount of data has been collected on the geographic and habitat distribution of pheasants, but their population biology, and the impacts of threats are still to be quantified. Translating research results into feasible conservation action involves science, sociology, economics and politics. The programs have to be monitored for their effectiveness. This is the current challenge in the area of pheasant conservation throughout Asia.

Forests and People

In traditional communities living along forests, the wildlife and forests are automatically conserved as the beliefs and lifestyle of such communities have been shaped to live in harmony with nature. It is only in the past several decades, that overpopulation and overexploitation of resources has caused an imbalance. Conservation programmes often have to involve the local communities which have been dependent on the forest resources, but have now relegated the traditional setup to the back seat. These programs invariably involve conservation education and awareness. On the other hand, development of roads, communication, healthcare, education, tourism infrastructure, and cottage industries will improve the standard of living in indigenous communities. Unless people have sufficient means to support themselves in a sustainable manner, natural resources will be used unsustainably for short-term gain, resulting in biodiversity loss. Supporting sustainable or alternative means of making a living is a universal activity for governments, conservation organizations and NGOs. This takes various forms in different countries – e.g. bee keeping, resin collection, collection of other minor forest products, medicinal plant cultivation, etc. The Indian Forest Rights Act of 2006, and the Aichi Biodiversity Targets 2011-2020, are examples of laws and guidelines that will help to achieve these goals (see 'Further Reading' at the back).

An example of sustainable livelihood is the harvesting of resins from the rain-forest Dipterocarp trees in Cambodia. As elucidated in a study titled 'Liquid resin tapping by local people in Phnom Samkos Wildlife Sanctuary', published in the *Cambodian Journal of Natural History* in 2009, government policy affects the way in which the people utilize the resources of the forests, and the government has to ensure sustainable harvesting of resources, and help to prevent alienation of villagers from the conservation process.

Liquid resin is a sticky substance exuded from the trees *Dipterocarpus costatus, D. alatus, D. dyeri, D. jourdainii, D. intricatus,* and other species, that usually functions to coat tree wounds or repel insects. The liquid resin called *chor teuk* in Khmer language is traditionally used for lighting fires, and waterproofing baskets and boats. Today, resin is mainly used in the manufacture of paint, vanishes and lacquers, as a fixative in perfumes, and to soak wood used to make floors, boats, and furniture. The resin is collected using various methods and sold locally, creating an income for communities in or on the edge of forest areas. Resin collection does not necessarily harm the trees, which can continue to be used by future generations (Neang, 2009).

In the 1980s, resin collection was the main source of income for many forest communities, especially minority groups, in the provinces of Kompong Thom, Preah Vihear, Mondulkiri, Ratanakiri, Kampong Speu and Pursat. During the 1980s and 1990s, resin trees in forest concessions contracted by the Cambodian Government were permitted to be selectively logged regardless of disagreement from villagers. After the disappearance of resin trees from these areas, the local people who had been harvesting resin turned to unsustainable activities, such as logging, and the collection of hard resin, mushrooms, rattan, vines and bamboo for their livelihoods. After the year 2000, when all forest concessions were suspended by the government, local communities living around the forest edges began to tap the scattered resin trees that remained in their areas to supplement their income. Today, Preah Vihear, Kampong Thom, Mondulkiri, and Oddar Meancheay Provinces are the main sources of resin exported from Cambodia (Neang, 2009).

People in the Phnom Samkos Wildlife Sanctuary (PSWS) in the Cardamom Mountains of Pursat Province, Southwest Cambodia, have traditionally depended on collecting various kinds of Non-Timber Forest Products (NTFPs). These resources have contributed to people's livelihoods in many ways through direct consumption, income-generation and as construction materials, medicines, ornaments and fragrances. The question of whether these resources have been harvested in a sustainable manner or not remains uncertain as the population in these areas has rapidly increased, putting pressure on the limited resources and possibly degrading natural habitats. The collection of liquid resin appears to have increased in PSWS since 2008, because of increased road access, increased market demand, higher prices offered by traders and tighter restrictions by ranger patrol teams on illegal alternatives. Of all the NTFPs surveyed in recent years, liquid resin has provided the most significant income to local tappers. After the zoning of natural resource management areas in the Phnom Samkos Wildlife Sanctuary was completed in 2007, all relevant stakeholders have been involved in a programme to protect and conserve resources in a sustainable manner. The zoning restricts access of local communities to resources in the Conservation Zones and Core Zones where more resin trees occur. However, local communities have the right of temporary ownership and access to NTFPs in areas designated as Community Protected Areas (CPAs) (Neang, 2009).

Evidently, conservation cannot succeed without local people's participation. This is especially true in the tribal areas where traditional dependence on forests has been high. In Arunachal Pradesh where large tracts of wilderness areas are outside PAs and legally protected forest categories, enlisting local support is *sine non-qua*. People's support will only be forthcoming if people see forests as source of sustained income to them. One of the activities that can generate substantial revenue without causing damage to the ecology of the area is eco-tourism. The state has very high potential for natural area based tourism, and if suitably planned, eco-tourism can become a major source of state revenue and local people's income. There have been recent initiatives in Eagle's Nest Wildlife Sanctuary in Arunachal Pradesh by an NGO called the Kaati Trust and the same model can be initiated in other parts of the state (Athreya, 2006). Also, neighboring country Bhutan is an example for thriving eco-tourism.

Ex Situ Conservation

What is *Ex Situ* Conservation? The definition of '*ex situ* conservation' according to the United Nations Convention on Biological Diversity (CBD) is – 'the conservation of components of biological diversity outside their natural habitats'. Ex situ habitats include zoos, botanic gardens, butterfly parks and aquaria (Maunder and Onnie, 2004).

As the extent of the world's natural habitat has reduced, the importance of the modern zoo to conservation is increasing. International studbooks, breeding programmes, and sophisticated techniques for genetic identification and management enable zoos to preserve species which would otherwise disappear. A more recent development is the use of zoo-bred animals to reintroduce populations into the wild, and thereby reinforce their small populations.

Although the *in situ* conservation of any species must always be the highest priority, in 2002, IUCN recognized that *in situ* conservation might not always be sufficient to ensure the long term existence of many endangered species of the world. The publication of the 'IUCN Technical Guidelines on the Management of Ex-situ populations for Conservation' (December 2002) has recognized that there are a range of complementary conservation approaches, including the use of *ex situ* techniques (Corder, 2007). The guidelines were drafted by a team established by CBSG (Conservation Breeding Specialist Group) at the CBSG Annual Meetings, and state - "the primary objective of maintaining *ex situ* populations is to help support the conservation of a threatened taxon, its genetic diversity, and its habitat" (Maunder and Onnie, 2004). In addition, WPA and IUCN have jointly published the 'Guidelines for the Re-introduction of Galliformes for Conservation Purposes' in year 2009, to help in the planning of such projects all over the world.

Ex situ facilities for wild species conservation - including zoos, botanic gardens, butterfly houses, insectaria, pheasantries, aquaria, and gene banks - represent a substantial investment, and are important means for retaining current levels of biological diversity. Since the original IUCN Policy Statement on Captive Breeding 1987, the science and practice of *ex situ* conservation has vastly improved.   A number of important changes, including new policy and legal instruments, have greatly altered the working context and objectives for *ex situ* conservation. There has been a fundamental shift towards using *ex situ* conservation for the conservation and recovery of wild populations. There has been a widespread increase in global experience with *ex situ* techniques, with the help of national and global networking of *ex situ* practitioners who are developing skills in species management. *Ex situ* conservation, and associated display and educational activities, is utilized as a tool to lever financial and scientific support for the conservation of important wildlife habitat areas. In addition we have seen major new institutional investments, notable examples include the Millenium Seed Bank, Royal Botanic Gardens, Kew (UK), and conservation biology laboratories at Kings Park and Botanic Garden, Perth (Australia), Smithsonian Institution (USA), and Chicago Botanic Garden (USA), providing new facilities and intellectual investment. 'Of course, not all zoos are able to contribute to conservation. The varying availability of technical and financial resources (and varying popular attitudes to wildlife) has resulted in a wide spectrum of institutions with vast differences in their standards of animal welfare, education and science (Maunder and Onnie, 2004). By and large it has been the European and North American zoos which have developed the science of the modern zoo. However, this pattern is changing with zoos in other regions such as those in Kuala Lumpur, Jakarta and Singapore, now becoming much more involved in local conservation and education. A major international initiative has been taking place to save the highly endangered Sumatran rhinoceros. Isolated individuals, found in 'doomed' areas where the forest is being cleared by loggers, were caught to establish breeding programmes both locally in Malaysia and Indonesia, and overseas in the United States and Great Britain. A similar joint initiative is being undertaken for the kouprey, which is also believed to be in imminent danger of extinction. The Kouprey Trust, involving six zoos, aims to conduct field surveys to locate the remaining populations and then to establish a captive breeding center in Vietnam, to build up their numbers. The goal is to maintain a secure reserve stock of these animals as a safeguard against extinction (Collins et al., 1991). Reintroduction projects for pheasant species are also being undertaken by WPA in Malaysia.

What Necessitates Captive Breeding?

Once a species' population has reduced below a certain number of individuals, through habitat loss, habitat degradation, or over-harvesting by humans, it tends to dwindle rapidly towards extinction, due to loss of genetic variability, and vulnerability to the fluctuating environmental variables. Anthropogenic disturbances have caused many wildlife species numbers to dwindle in all types of ecosystems. As a result, the conservation of small populations is one of the most challenging tasks of our age. It requires the combined effort of the science of conservation ecology, genetics and wildlife management, as well as policy support from the governments (Dhiman, 2010).

The absolute need for *ex situ* capacity has been justified and demonstrated through successful captive breeding and reintroduction projects that have ultimately established new populations of some species (Maunder and Onnie, 2004). Reintroduction into suitable habitats is probably the best option to rapidly counteract immediate extinction risks. Habitat protection and reintroductions have facilitated the successful recovery of ungulate mammals, such as the alpine ibex (*Capra ibex),* the Arabian oryx (*Oryx leucoryx*), and the bighorn sheep (*Ovis canidensis*). *Capra ibex* was reduced to fewer than 200 individuals in the 18th century in Europe due to overhunting, but reintroductions have increased the population to about 40,000 individuals. Eventually, providing wildlife corridors to connect the isolated habitats (that are currently fragmented by human activities) will help the species expand into a network of small populations. In many threatened species, it is important to combine environmental and genetic studies. Studies to investigate genetic variation (using microsatellites and mitochondrial DNA control region sequences) are being used for designing management plans for species whose habitat has been severely fragmented.

Conservation Breeding in Indian Zoos

India is among the 12 mega biodiversity countries of the world. The country has immense natural wealth, despite problems of habitat loss, degradation and fragmentation due to anthropogenic pressures (Central Zoo Authority).

Forest legislation in India dates back to 1865 when the first Indian Forest Act was passed. This Act was later revised in 1878. The process of revision continued till the currently applicable Indian Forest Act of 1927 was formed. The Wild Life (Protection) Act, 1972 was enacted to give proper shape to wildlife conservation in the country. The Forest (Conservation) Act was enacted in 1980 with a view to check indiscriminate diversion of forest land for non-forestry purposes. The Central Zoo Authority was created by the Government of India in the year 1992 through an amendment of the Wild Life (Protection) Act, 1972. Its main objective was to enforce minimum standards and norms for upkeep and healthcare of animals in the Indian zoos (Central Zoo Authority).

Strategy for Captive Breeding

To give proper direction and thrust to the management of zoos in the country, the National Zoo Policy was framed and adopted by the Government of India in the year 1998. The main objective of the zoos under the National Zoo Policy is to complement and strengthen the national efforts in wildlife conservation, through education and research. There is a need to infuse more technical and scientific culture in operation of our zoos and change the general perception of zoos from mere picnic spots to more of a scientific institution (Central Zoo Authority).

The National Wildlife Action Plan lays emphasis on the role of zoos for *ex-situ* breeding of endangered species of wild fauna, and their rehabilitation in the wild. The Central Zoo Authority has been identified under the Plan, as one of the organizations for developing capabilities in this field by ensuring healthy captive stocks for display, and genetically healthy animals for reintroduction in the wild. About 35 animal species (mammals, birds and reptiles) were identified for 'planned coordinated breeding program' in Indian zoos. The responsibility for maintaining the studbooks for selected endangered species was given to zoos (Central Zoo Authority). It is interesting to note that zoos can exhibit animals that have been born in zoos, or exchanged between zoos, or have been rescued, but capture of animals from the wild for display in zoos is not permitted in India.

Fig 4.10 A signboard at the entrance of the Himalayan Nature Park in Kufri, Himachal Pradesh, India. Photo by Anita Chauhan 2009.

All the zoos in India are equipped with small veterinary facilities along with veterinary personnel as per the classification of the zoo. Indian Veterinary Research Institute, Bareilly (Uttar Pradesh) has been identified as a National Referral Centre (NRC) to provide super specialty services and diagnostic facilities to the Indian zoos to deal with the issue of health care of wild animals, training of zoo veterinarians, and conducting research on health care and nutrition of wild animals in captivity. Another such facility is available at the 'Wildlife Health Monitoring, Disease Diagnostic and Research Cell' at Veterinary College, Jabalpur, Madhya Pradesh (Central Zoo Authority).

The coordinating and participating zoos of the coordinated breeding program have been asked to construct appropriate enclosure for the targeted wild animal species, to fulfill their physical and behavioral needs. The coordinating zoos for each targeted species have also been requested to create off-display conservation breeding facility either in the zoo compound, or as satellite facilities. School of Planning & Architecture (SPA), New Delhi, has been assigned the study on Zoo Design and Architecture to help the zoos in this regard. In order to infuse new technology in the field of reproduction and molecular characterization of endangered species, a Laboratory (Laboratory for Conservation of Endangered Species – LaCONES) has been established at Hyderabad. A Memorandum of Understanding (MoU) has been signed with the Wildlife Institute of India, Dehradun for preparation and updating of National Studbook for the identified endangered wild animal species being taken up for conservation breeding program. The zoos are also being provided funds in the form of small grant fellowships to organize studies to deal with the local issues (Central Zoo Authority).

Fig 4.11 A male western tragopan in the circular pheasantry at Kufri Nature Park, Kufri, Himachal Pradesh, India. This male was brought from the pheasantry in Sarahan, H.P., where some success has been achieved in reproducing this species. Photo by Anita Chauhan, 2009.

Another component of the program is the identification of a Protected Area having wild population of the proposed species or, a re-introduction site in the vicinity of the conservation breeding facility. The in-situ managers of the protected areas will be taking corrective measures to address the cause of decline/ extinction of wild population of the targeted species in its natural habitat. More than 90% of the recognized Zoos in the country are operated or controlled by the State Forest/ Wildlife Departments. These Departments are also managing the in-situ facilities which make the coordination between the in-situ and ex-situ wildlife conservation activities much easier.

The help of the national/ international organization, institutions, and NGOs will also be sought to make the programme successful. World Association of Zoos & Aquariums (WAZA) will also be requested to support the activity as a part of global species management programme. Conservation Breeding Specialist Group of SSC-IUCN will also be engaged in the activity. The wild animals bred as part of the coordinated conservation breeding activity, will occasionally be released in the identified habitats following IUCN guidelines (Central Zoo Authority).

Fig 4.12 A. New conservation breeding enclosures at Khariun, Chail, B. Cheer pheasant at Khariun pheasantry. Chail, Himachal Pradesh, India. Photos by Satpal Dhiman.

Table 4.3: List of the identified pheasant species to be taken up for planned coordinated conservation breeding, with the coordinating zoos, participating zoos, and the number of birds in captivity in India.

Name of the Species	Name of the coordinating Zoo	Names of the participating Zoos	Number of animals in captivity
Himalayan monal (*Lophophorus impejanus*)	Manali	Darjeeling, Gangtok	23
Blood pheasant (*Ithaginis cruentus*)	Gangtok	Darjeeling	--
Cheer pheasant (*Catreus wallichii*)	Chail	Almora	48
Hume's pheasant (*Syrmaticus humiae*)	Aizawl	--	4
Grey Peacock pheasant (*Polyplectron bicalcaratum*)	Guwahati	Kolkata, Darjeeling	60

Sclater's monal (*Lophophorus sclateri*)	Yachuli	--	--
Tibetan eared pheasant (Crossoptilon harmani)	Yachuli	--	--
Temminck tragopan (*Tragopan temminckii*)	Yachuli	--	--
Blyth's tragopan (*Tragopan blythii*)	Kohima	--	12
Western tragopan (*Tragopan melanocephalus*)	Sarahan	--	19
Satyr tragopan (*Tragopan satyra*)	Darjeeling	Gangtok	2
Grey jungle fowl (*Gallus sonnerati*)	Tirupati	--	33
Red jungle fowl (*Gallus gallus*)	Morni	Chail, New Delhi, Aizwal	209

What are Studbooks?

Initially, studbooks were used to keep a written record of the pedigree of a purebred stock, especially of racehorses and dogs, for breeding and trade purposes. A studbook is literally a register in which the origin (descent) and characteristics, of the registered animals of one species, are recorded. A studbook can be used for the management of a species in captivity in zoos/aviaries, and countering inbreeding by working with breeding programs. Each studbook has one or more 'studbook keepers' who are responsible for maintaining suitable records, and guiding the program (Bocxstaele, 1992).

Studbooks contain the registration number of each animal of the particular species in captivity, its sex and birth-date, the identity (registration numbers) of its parents, where it was born and where (and when) it was transferred to, ownership information, as well as its house name, and its identifiers (such as transponders, tattoos and tags). The update section of the studbook compiles any births, captures, transfers, deaths, and releases during the reporting period (Althaus et al., 2010). The goal of this is to guarantee the genetic health of the population in the long term. Animals and their offspring can also be exchanged between studbook members under a co-operative breeding-loan agreement, with the studbook keeper possibly playing a mediating / advising role regarding the husbandry and breeding of the species (Anonymous).

For species that are in danger of extinction (in captivity and/or in the wild), it is vitally important to maintain a healthy gene pool of the population. The smaller the population, bigger are the risks of inbreeding. Indeed, a study on various captive deer species has suggested that individuals from those species managed by an international studbook had a higher relative life expectancy (Althaus et al., 2010). Co-ordination is of utmost importance for the success of breeding programs. Not only should all the different types of research and reports be co-ordinated in order to create a healthy viable population of captive birds, but even a simple census, a list of all individuals of a species in captivity is essential. The WPA is also a part of the Captive Breeding Specialist Group for Galliformes, and maintains several international and European studbooks for pheasants (Bocxstaele, 1992).

As mentioned before, the WPA was started by aviculturists in Europe in 1975 because they were facing the problem of inbreeding in the pheasants, since it was no longer possible to get a steady fresh supply of pheasants from Asia; and it became a means to replenish the dwindling pheasant populations in their native habitat in Southeast Asia. Two of the first studbooks to be started were for the Edward's pheasant by WPA (1975), and for the white eared pheasant by Jersey Wildlife Preservation Trust, UK (1986). Registers and studbooks originally were a simple list. Parent-registration was the first tool to be used to avoid inbreeding. Today, a studbook is much more than just a registration of individuals and therefore requires a team of specialists. The final co-ordination and evaluation of the results remain the job of one man, the studbook-keeper. He/she can be assisted by regional studbook-keepers who can be responsible for sub-populations of captive pheasants on each continent.

Studbooks maintained by WPA include –

International studbooks for Blyth's tragopan and Cabot's tragopan.

European studbooks for Malay peacock pheasant, mountain peacock pheasant and great argus pheasant.

International studbooks kept by others –

- Congo peafowl – Royal Zoological Society of Antwerp, Belgium.

-Mountain peacock pheasant – Bronx zoo, NY, USA and Dept. of Wildlife and National Parks, Malaysia.

- Malayan peacock pheasant – Bronx zoo, NY, USA.

There are also national studbooks kept by various countries. In India, the Central Zoo Authority maintains the National studbooks for 6 species of pheasants. Besides these, there are many local studbooks for various pheasant species kept in zoos in India. Some of the studbooks give detailed information such as the 'Biology and Status of the species' (behavior, distribution, Red List status, threats), 'Population Planning' and 'Recommendations', 'Demographic and Genetic Analyses' of the captive population, and 'Pedigree Charts', in addition to the 'Census List'.

Captive Breeding of Pheasants

Captive breeding of pheasants requires much care and expertise. A lot of literature is available about the aviculture (including 'Feeding and Nutrition', 'Stock and Breeding', 'Incubation, Hatching and Rearing', and 'Diseases') of pheasant species. The captive bred pheasant population in India is only found in the government owned zoos and pheasantries.

Captive breeding programs must address the issue of adequately preparing animals behaviorally for life in a wild environment . This is an especially formidable task with animals that have a complex social system, and whose behaviors for mating, communication, foraging, predator avoidance, parenting, and migration are learned by observation of the parents or other experienced individuals. A captive environment does not adequately replicate natural conditions, or ensure that exposure to appropriate learning opportunities occurs.

Some Galliformes species breed readily in captivity, although the rarer the species, the more difficult it is to captive-breed. Successful breeding might involve artificial insemination, vaccination, artificial incubation of eggs, the use of broody chickens, or natural rearing by the pheasants themselves. Unless captive breeding is planned in such a way as to encourage natural characteristics, the young birds that are bred, often compare very poorly to their ancestors. The keeper has to design aviaries that encourage natural behavior, and to provide an environment where the birds feel sufficiently at ease to raise their young and teach them valuable life skills. Over the years, practical experience accumulated from the WPA has resulted in a number of re-introduction projects for pheasants in Asia (Corder, 2007).

Although it is possible to advocate some general guidelines for conservation breeding, and for re-introductions, specific protocols for each species need to be developed. For example, the peacock pheasants lay a clutch of just one or two eggs, and the hen puts in a great amount of effort to ensure the survival of her young, whereas the ring-necked pheasant may hatch more than a dozen eggs at a time. Most pheasant species are reared by just the female, who also does all of the incubation, whereas both the male and female cheer pheasant are involved in raising their young. Tragopans are unusual pheasants in that they nest in trees and are very arboreal. In captivity, they are provided with perches for them to roost on at night. However, these aviary perches seldom offer the bird an opportunity to conceal itself within the foliage, as its wild relatives would do. Thus, if a tragopan is released from captivity as part of a re-introduction programme, it will almost certainly roost on an open perch and fall victim to the first predator that comes along (Corder, 2007). Also, the young should be allowed to overwinter with their parents to learn natural behaviors.

A.O. Hume has, in his book 'The Game Birds of India, Burmah and Ceylon Vol.1 (published in 1879)', described this problem w.r.t. the grey peacock pheasant *Polyplectron bicalcaratum* –

"We are told that when the young of this species were first hatched in the Zoological Gardens (England), a Bantam hen was employed as a foster mother, and that the chicks would follow close behind her, never coming in front to take food, so that, in scratching the ground, she frequently struck them (inadvertently) with her feet. The reason for the young keeping in her rear was not understood until, on a subsequent occasion, two chicks were reared by a *Polyplectrum* hen (natural mother), when it was observed that they always kept in the same manner close behind the mother, who held her tail widely spread, thus completely covering them; and there they continually remained out of sight, only running forward when called by the hen to pick up some food she had found, and then immediately retreating to their shelter (under the tail)."

Conservation breeding projects for cheer pheasant *Catreus wallichii* and western tragopan *Tragopan melanocephalus* in the state of Himachal Pradesh, India are described in chapter 6.

Reintroduction

In re-introduction programmes, the scientists have to ensure that suitable habitat is available for re-introduction, and that the reasons for the birds' previous extirpation no longer exist. A 'Communities and Education Programme' involving the local people who share the natural resources might be very necessary before and during any reintroduction.

After release, captive-reared animals must be monitored to determine whether they have been able to survive the stresses of living in a wild habitat. To ease the transition from captivity to the wild, the release may be somewhat gradual. For example, a 'soft release' may involve providing a large net enclosure in the natural habitat for acclimatisation, or the provision of food at the release point until animals learn to forage on their own. Monitoring the released population is necessary for the assessment of survival and the causes of mortality, so that future releases can attempt to avoid such pitfalls.

Programs of captive breeding and release can be extremely expensive, and their success may be limited because of difficulties in biology, ecology, and in addressing the ultimate cause of the species decline (such as habitat loss, or excessive hunting). Reintroduction efforts should always be accompanied by a program of public education.

Wildlife reintroductions can help to restore populations and save species from extinction. 'Reintroduction will, however, remain one of the last courses of action to be taken, because of the resources needed, and the sort of detail on the species' ecological and behavioral requirements that are necessary to ensure that the reintroduction attempts are successful' (McGowan, 1994).

Success rates of reintroductions of captive-bred birds are low due to behavioural problems that result in high rates of predation after they are released. In particular, the released birds may use habitats maladaptively, leading to an ecological trap, i.e. preference for low-quality habitats leading to reduced survival and/or breeding success. Ecological traps in reintroductions can be identified only through intensive studies of habitat preferences and survival of individuals, but such studies are lacking for most species (Rantanen et al., 2010).

For example, in a study titled 'Habitat preferences and survival in wildlife reintroductions: An ecological trap in reintroduced grey partridges' published recently in the *Journal of Applied Ecology*, scientists have investigated habitat preferences and their relationship with survival, by radio-tracking reintroduced, captive-bred grey partridges *Perdix perdix* (a commercial game and native farmland bird of conservation concern in the UK). The low success rate of released grey partridges could be due to maladaptive habitat use. As a result of the study, in grey partridge reintroductions, scientists have recommended releasing grey partridge family groups in autumn rather than releasing pairs of birds in spring, and providing game-covers (areas of tall vegetation specifically planted to provide cover for game birds) that could induce the groups to settle onto the release areas (Rantanen et al., 2010).

In reintroductions in general, the habitat preferences of the released animals should be documented together with their outcomes, to enable the detection of possible ecological traps (Rantanen et al., 2010). The results of such studies are useful for planning reintroductions of other Galliformes, including the pheasants.

In situ methods of habitat management, research, and monitoring of populations, as well as *ex situ* methods of captive breeding in zoos and conservation breeding facilities, are helping to recover degraded habitats and revive wildlife populations.

Chapter 5

Field Notes

I made my way slowly upstream, creeping over the great rounded boulders, or wading through the rush of icy water. Every turn revealed new beauties. An enormous overhanging mass of quartz loomed up draped with swaying vines, and beyond, a little sandy bay was fretted with the tracks of pheasants, cats, and deer. In the spots of sunlight among the higher branches, crimson butterflies flitted about, and white-fronted red-starts dashed ahead from stone to stone.

From *Pheasant Jungles* by William Beebe

Fig 5 A male green junglefowl from *A Hand-book to the Game-birds* Vol. I and II, by W.R. Ogilvie-Grant, 1897.

Chapter 5

Field Notes

Stopping at a favorable opening, a half-mile up stream, I began my laborious climb upward, first through a steep ascent of soft mold densely shaded by wild bananas. The undergrowth seemed scant, and as I brushed aside the first thicket of soft-leaved plants, I anticipated an easy first stage. But the gray down on the myriad green stems proved scourging whips of nettle which lashed face and hands at every step. There was no alternative, so I clambered painfully on, seizing hold of every cold, smooth-enamelled banana trunk as a haven from the merciless needles.

A small side ravine spread out into a broad, fern-filled bog, and the nettles were left behind. Then came more bananas and small evergreen trees with little or no under growth. Here was the feeding ground of the pheasants and deer. There was hardly a square yard of mold which did not bear the marks of the tiny hoofs of the barking deer or the strong claws of the birds. Now and then I picked up a feather of some silver pheasant clinging to a bramble on the steep slope.

Two feathers caught my eye. They were from the plumage of no silver pheasant, but brilliant, iridescent, changeable green and purple. I was at a loss to know from what gallinaceous bird they had come. A little way farther I found another. Later, while worming my way through a barking deer's tunnel at the roots of a perfect tangle of bamboo, I heard subdued chuckles and the rustling of leaves ahead. A few feet brought me to a deeply worn but steep sambur trail, along which I made my way on hands and knees, without making a sound.

The rustling of leaves and the spray of earthen pellets falling down, came more distinctly to my ears, and at last I rested for many minutes with my face buried in a clump of blue, sweet-scented pea flowers.

Inch by inch I then edged myself upward, digging with fingers and toes into every deepened hoof-rut. A shower of earth fell upon me, and with joy I saw that a clump of soft-leaved, mint-like plants lay before me. I did not have to increase my numerous wounds by a slow penetration of either nettles or briers.

The revelation came sooner than I expected. Noiselessly plucking away leaves and stems one by one, to form a low tunnel, I pushed slowly and cautiously ahead. Never have I been "closer to Nature" than on this trail. My trail was more like that of a snail or worm than of any vertebrate! Glints of light filtered through the green ahead, and I saw that a low, perpendicular bank of earth barred my way on each side. Then the forms of 1 or 2 birds appeared, and with a screen of leaves still intervening, I watched what was probably the first wild Sclater's impeyan (monal) ever seen by a white man.

-   From *Pheasant Jungles* by William Beebe

The following articles illustrate the painstaking research work done during the past 15 years or so, which has provided vital data necessary for understanding and conserving the pheasant species. The excerpts are from the former newsletter of the World Pheasant Association-UK, *Tragopan* (from 1995 to 2003), and have been duly edited for the general readers. This newsletter and the new WPA newsletter, called *G@llinformed*, are available at the following website: http://www.galliformes-sg.org/pheamain.html .

1.   THE MATING SYSTEM AND BREEDING ECOLOGY OF CABOT'S TRAGOPAN *TRAGOPAN CABOTI*

Eleven Cabot's tragopans were radio-tracked in the Wuyanling Nature Reserve, Zhejiang Province (China) from November 1990 to June 1992. The mating system and breeding ecology were studied. The results are as follows:

1.   The line-transect method and Fourier series estimator basis on perpendicular distance were used. The population density was 14.65 ± 6.77 birds per square kilometer in autumn and there were about 90 ± 41 tragopans in the reserve.

2.   The monthly home range area of the radio-tagged birds was 1.74-20.70 ha. The change of the home range position and area was related to the birds' breeding activities, seasons, habitat types and food abundance.

i)   The home range area in early spring was significantly larger than in winter: the dominant cock had a stable home range. The hens wandered over a large area and often went to the home range of, and fed together with the dominant cocks. The enlargement of home range was closely related to the onset of the breeding season.

ii)   The hens had the largest home ranges in the pre-breeding period. Home range size decreased whilst hens were incubating eggs and increased again during the hatching period.

iii)   The range areas were closely related to the type of forest, and were larger in the mixed coniferous forest.

3.   The preferred habitat types in different seasons were: evergreen-deciduous forest and broadleaf strips in spring; broadleaf strips and shrubs in summer; evergreen-deciduous forest and shrubs in autumn; the habitat in winter was the same as that in spring. The habitat characteristics were significantly different between seasons.

4.   The dominant cock and a hen formed a monogamous pair in the early breeding period, the immature birds and the non-dominant males did not breed. The mating relationship showed three forms in the late breeding period.

i)   The dominant cock stayed near the nest during the early period of incubation and then wandered and fed alone or with other males.

ii)   The dominant cock held a stable home range and mated with another hen which came into his home range, while his first mate began incubating eggs. This was interpreted as resource defence polygyny.

iii)   The hen could search for another cock and breed again if her first clutch was damaged in the early incubating period.

5.   Fifteen nests were found in this study and about 86.7% of them were located in the *Cryptomeria* woodland. 80% of the nesting trees were *C. fortunei*. The compositional analysis indicated that there were four factors which affected nest site selection: availability of nesting trees, terrain, elevation and distance from the hill ridge.

6.   Only 15.8% of the nests were incubated successfully. The loss of the nests and eggs is believed to be one of the main factors causing Cabot's tragopan to be endangered.

7.   The chicks were brooded by the female alone and the made birds did not participate in brood rearing. No obvious altitudinal movement was detected during this period.

[This thesis abstract is re-published form WPA-China News 5, issued in June 1994].

Ding Chang-qing, Department of Biology,

Beijing Normal University, Beijing 100875, PR China

2.   FIELDWORK CONDITIONS WHILE STUDYING CABOT'S TRAGOPAN

The Cabot's tragopan *Tragopan caboti* is a threatened species endemic to the lower montane zone of south-eastern China. In the past ten years, a series of researches on the pheasant have been conducted in the Wuyanling Natural Reserve, an offshoot of the Wuyi mountain range, where the type specimen of Cabot's tragopan was collected in 1857.

Wuyanling Natural Reserve is located in Taishun country, Zhejiang province, about 1,800 km south of Beijing. Because of the great distance and transport difficulties, the journey there is quite long. It takes 28 hours from Beijing to Hangzhou by train, then 15 hours to Taishun county town by coach, and finally more than four hours on a mountain bus trip to Wuyanling which is only 60 km away from the county town.

The total area of the reserve is 1495 ha consisting of two parts. The central core area for wildlife conservation and especially for Cabot's tragopan, lies between 700 m and 1500 m elevation and the main vegetation is subtropical mixed evergreen-deciduous forest.

Surrounding this there are commercial conifer plantations which have replaced the primeval evergreen-deciduous forests. These forests were felled in around 1970, and commercial conifers, such as *Cryptomeria* were planted. However, several broadleaf strips, about 20-50 m wide, remain along the streams.

Fig 5.1 A male Cabot's tragopan from 'A Hand-book to the Game-birds' Vol. I and II, by W.R. Ogilvie-Grant, 1897.

The Cabot's tragopans are distributed in the upper part of the reserve. Their typical habitat is the mixed evergreen deciduous-coniferous forest, dominated by the plant family Fagaceae. In recent years, according to our radio-tracking study, the tragopans can also live in the broadleaf strips and the hens prefer nesting in *Cryptomeria* trees. This shows that Cabot's tragopans can adapt to changed habitats.

The living conditions in Wuyanling were very hard. It seemed to always be raining which made it very cold and damp. Moreover it was no warmer in our room, the temperature being almost the same inside as in the open air. The food was simple with no fresh vegetables except bamboo shoots and a kind of fern. Although the bamboo shoots and fresh ferns are delicious dishes and very expensive in big cities, because they were our main food, we were always hungry and had no power to climb the mountains.

In order to hear the calls of the tragopans and to observe their behavior, we often got up early and set out before sunrise. The morning dew made our clothes so wet they felt like a shirt made of ice. Around 0700-0800 hr, a local assistant arrived and brought me breakfast. Usually it was a large mess-tin of noodles which were still warm. I felt it was the most delicious meal of the day.

During the daytime we were busy doing our fieldwork, such as radio-tracking, censusing populations, observing behaviors and studying habitat samples. Sometimes we would bring some light food with us for lunch when we had a long walk from the base camp.

Generally, we came back after sunset after the tragopans had gone to roost. After supper, I looked through the field notes, made calculations and put the location points of the radio-tagged birds on the map, summarized the whole day's results and thought what to do the next day. Since there was a shortage of electricity in the reserve, all had to be done by candle-light.

Loss of the nest and eggs is one the key factors causing Cabot's tragopan to be endangered. In the breeding season, we made a small hut in the forest, and guarded the nests all day long. In the earlier years of the research, Professor Zheng and Zheng-wang (Chairman and Secretary of WPA-China from Beijing Normal University) used to sleep in the open all night in order to protect the nests from damage by small mammals such as *Charronai flavigula* and *Felis bengalensis*.

Although the fieldwork conditions were very hard and full of difficulties, I think it was quite a rare chance for me to gain useful experience. It benefited me a great deal in my future research work.

In China, most of the pheasants are distributed in the remote mountain areas, which are mostly undeveloped and where living conditions are harsh. Almost everybody who has worked on pheasants has had similar experiences. I was lucky because Professor Zheng's research group has made a very good friendship with the people of the reserve for more than ten years. The people of Wuyanling often looked after me well and treated me as one of their own family members. They helped me overcome most of the difficulties, gave me a relatively comfortable room to live in, and served me warm food. However, other Chinese fieldworkers might not have these advantages. They have to sleep outside in tents waging struggles against the dark, cold and dampness, and threat of wild animals. They may make their own food or eat cold food. Gastric diseases are quite common among Chinese fieldworkers.

At last, I want to express my thanks here to the people of Wuyanling Natural Reserve and those who helped me during my eighteen months of fieldwork.

Ding Chang-qing, Institute of Zoology, Academa Sinica,

Beijing 100090, Peoples' Republic of China

3. CAPTIVE BREEDING INITIATIVES FOR THE VIETNAMESE *LOPHURA* SPECIES

Vietnam is in the process of opening its doors for industry, tourism, and conservation, and the traditional scene is quickly changing. The people of Vietnam are recovering from the wars that devastated their beautiful country, and are working hard and with great optimism. They are eager to learn methods developed in other parts of the world in the past decades, when Vietnamese people only had time to heal the wounds of the past.

Vietnam is comparable in size with Norway, but with 20 million people to feed and provide jobs for. Foreign botanists and zoologists are becoming increasingly interested in this beautiful country and have, on the one hand, discovered that there are still new species to be found, but on the other, that some species are nearing extinction. Many projects have been implemented to assist the Vietnamese in their endeavors to save and preserve their last pockets of virgin forest. Their admired chairman, Ho Chi Minh, did not take away conservation from the agenda, even during tremendous pressure throughout the Vietnam War, and established a national park close to Hanoi. However, it is clear that the huge population is now in competition with the last remaining wild places. The Government, and conservation organizations, are trying to find a way to find the balance between the needs of the people and the needs the wild fauna and flora of Vietnam.

Two recent events have prompted a cooperative attempt to relieve pressure on wild populations of threatened pheasants in particular. These became known at the 'Pheasants in Asia' Symposium in Lahore, Pakistan in 1992 and are:

- Nguyen Cu and Jonathan Eames of BirdLife Vietnam have not found any evidence that Edwards' pheasant *Lophura edwardsi* still exists in its known distribution area. Hardly any forest is left and this endemic species may be extinct in the wild.
- Hanoi Zoological Garden kept and bred the Vietnamese pheasant *Lophura hatinhensis*. Mr. Le Si Thuo, the Director, and Mr. Dang Gia Tung, Curator, have achieved remarkable success with this species and numbers have increased to over 30 individuals. The Director intends to send pairs on breeding loan to approved collections outside Vietnam under the auspices of the World Pheasant Association, in order to avoid casualties by disease. The birds will, however, always remain the property of the Hanoi Zoo and the people of Vietnam.

This project intends to improve conditions in both of the Vietnamese Zoos, and to establish viable populations so that there is no need to take any more birds from the wild. Unfortunately, many birds are still taken from the wild, and we are working towards eliminating this threat to the species' survival. The taxonomic relationship between the two target *Lophura* species is still not clear.

Mr. Dang Gia Tung presented an excellent paper on the breeding and care of *Lophura hatinhensis* during the 1st International Captive Breeding Symposium in 1993 in Antwerp Zoo. The following year, agreements were signed between the World Pheasant Association and the Hanoi Zoo to establish close co-operation over a period of at least 5 years. The main items within the agreement are - training, establishing studbooks, and education.

In 1995, Gillian Stewart, formerly of Dublin Zoo in Ireland, and with experience in Pakistan and India, went to Hanoi and Saigon Zoos to provide technical assistance and establish a computerized record-keeping system. The software and hardware necessary were provided by the World Pheasant Association, as was a fax machine to improve direct communication between the zoos and the WPA. Most of the work undertaken, and recommendations that resulted, are given in a series of reports. In due course, these will be the basis for an action plan for captive breeding and education activities in Vietnamese Zoological Gardens.

One of the more important uses of these reports is the establishment of an Indo-Chinese Fauna Interest Group. This broad support group will be initiated by WPA and some Zoological Gardens in Europe, and will work in close co-operation with the Conservation Breeding Specialist Group of IUCN-The World Conservation Union. We hope that this Interest Group will prevent approaches, by zoos and individuals who visit Vietnam in order to obtain animals for personal gain. A second aim is to increase the financial support for joint projects taking place in Vietnam. WPA, Poznam Zoo, Muenster Zoo, Parc Zoologique de Cleres, London Zoo, and the Zoological Society for the Conservation of Species and Populations are forming a core committee for this initiative.

The possibly extinct Edward's Pheasant was established in captivity in Vietnam recently, as no Edward's Pheasants were in any public collection in the country. The WPA donated and transported 4 pairs of Edward's Pheasant to the Hanoi Zoo. The Zoo and WPA received a very good press, and this event was even featured on television in Vietnam. Dang Gia Tung was able to breed the species in 1995, in the year after the birds arrived from Europe, which is an achievement in itself. Six Edward's Pheasants were raised. Hopefully this species, which we can be sure only exists in captivity, will help in pheasant conservation world-wide, by acting as a very strong reminder of how easily a species can be lost in the world forever.

Hanoi Zoological Garden's best facility is the partly Brehm-Fund sponsored pheasantry complex, in which there is considerable breeding success. If over-crowding, due to lack of space, could be eliminated, even better results might be achieved. However, the zoo is in the process of developing the first breeding centre in the country, especially for Vietnamese species, and with emphasis on *Lophura* species.

The nominate subspecies of silver pheasant *Lophura nycthemera nycthemera* is regularly bred, and *L.n. beaulieui* and *L.n. annamensis* have not yet been bred but may be in the near future. *L. diardi* occurs in Vietnam, but is not endemic, and is kept in both Zoos.

In Saigon Zoo in Ho Chi Minh City, the situation is similar to that in Hanoi. WPA has signed agreements and will further extend the co-operation with this zoo in the coming years. Saigon Zoo also has good accommodation for Galliform birds, and further accommodation for pheasants is planned. Over-crowding does limit breeding potential and will eventually cause casualties. Presently *L.n. annamensis*, *L.n. nycthemera,* and also *L.diardi,* are kept.

In close co-operation with the Saigon and Hanoi Zoos, preservation programmes are being started for the Vietnamese pheasant *L. hatinhensis*, Edward's pheasant *L. edwardsi*, and Annamese silver pheasant *L.n annamensis*. Alain Hennache from France, Eddy Powell from the UK, and Dang Gia Tung and myself, are working together to establish and maintain an international studbook for Edward's pheasant, and towards inclusion in the European Endangered Species Programme (EEP).

In conclusion, the breeding of *Lophura* species in Vietnamese Zoos is developing quickly. Record-keeping, incubation techniques, and additional training of staff, will further improve the breeding for conservation of this group of pheasants. A workshop on Vietnmese galliform species will certainly further help the development of awareness for conservation in the country. A workshop is planned in early 1996. A detailed action plan based on reports of Gillian Steward will be available in a few months.

Han Assink, Captive Breeding Advisory Committee, World Pheasant Association, Netherlands

4.   SWINHOE'S PHEASANT IN YUSHAN NATIONAL PARK, CHINA

In Yushan National Park, hidden in the beautiful, ancient broadleaf forests, there is an equally beautiful and ancient pheasant, found nowhere else on earth.

Twenty to thirty years ago, when hunting and trapping pressures on the pheasants were high, and when replacing broadleaf forests with coniferous plantation was common practice, the Swinhoe's pheasant *Lophura swinhoii* faced severe threats to its survival in the wild. It was extremely difficult to see a Swinhoe's pheasant in the wild.

Today, Yushan National Park provides the Swinhoe's pheasant with a safe haven. Within the park, all one has to do to see a Swinhoe's pheasant, is go to the right habitat, become a part of the tranquility of its environment, and be alert to all the signs of life in the vicinity. Getting up before dawn and walking very quietly along a trail in the broadleaf forest may be all that is required of a visitor. Of course, in these cases one may only catch a fleeting glimpse of this secretive bird. To see and to learn about this species takes a lot more time and effort than that.

Swinhoe's pheasant is a large chicken-like bird. The female is reddish brown with tan-coloured arrows scattered throughout its body; the arrowheads are broader and more intricately patterned on the chest while more condensed and bright on the wings. The bare skin on her face is red. Her outer tail feathers are chestnut. She can be most easily distinguished from female Mikado pheasant *Syrmaticus mikado* (also found in the park), by her red legs.

The glorious male Swinhoe's pheasant is clothed in navy blue. The edge of each blue feather glistens in metallic hues in the right light. His crown, upper back, and the two central tail feathers are pure white, with maroon shoulders, while the edges of his wing coverts reflect a greenish blue. The red flesh of his face can expand into long erect horns and extended lobes during courtship. His legs are also red.

Swinhoe's pheasants are taller and slightly larger than Mikado pheasants. The male is about 72cm from head to tail, while the female is about 55cm.

Hunters hired by Robert Swinhoe, a diplomat naturalist from Great Britain, brought to him in Tamsui a pair of Swinhoe's pheasant in April 1862. This was when the scientific community discovered this species. By 1903, this pheasant was common in European aviaries. John Gould, who first described this species scientifically, was very impressed with its beauty. Among all the new species Gould described, he thought this species the most suited to be named after Robert Swinhoe who contributed so tremendously to the knowledge of oriental birds. Robert Swinhoe described this species as a true jungle bird. Its habitat was described as dense forest, or thick and dense forest, by various naturalists in the past. Nowadays most people catch sight of this pheasant along logging roads or mountain trails, which could be open, or lined by vegetation of various heights and densities.

Originally, Swinhoe's pheasants were found from almost sea level up to 2500 m in elevation. The main concentration of this species was in the Central Mountain Range. Swinhoe's pheasants appeared mostly in primary broadleaf forest, in areas where the canopy was closed, the undergrowth sparse, and the slopes gentle. Such places could be park-like where sun light filters through the canopy producing shifting dark and light patterns on the leaf litter on the forest floor. They also used mature secondary broadleaf forest where the canopy was partially open with moderate undergrowth. Some birds were also seen in mixed forest, or mixed bamboo forest with dense ground cover.

At present, much of the suitable lowland habitat in Taiwan has disappeared. Very little of this region remains undisturbed by human activities. Except in a few protected areas, most of the recent sightings of Swinhoe's pheasants are in deeper mountains above 1000 m. Between 1000-2000 m, large tracts of the original forest have already been converted to plantations, changed to other uses, or otherwise fragmented. Loss of habitat continues to pose the major threat to the future of Swinhoe's pheasant.

Swinhoe's pheasant is a very wary bird. It tends to avoid direct sunlight and is, therefore, more frequently seen during twilight hours. It is more frequently seen along logging roads or trails because it is easier to detect there than in the forest interior. It often walks slowly, pecking at low vegetation or the ground, as it moves forward. When the sun is up, when the roads or trails are brightly lit, the Swinhoe's pheasant generally retreats into the shadows of the forest. Interestingly, even the brightly coloured male Swinhoe's pheasant is remarkably difficult to detect in a forest, probably because the contrasting bright sunny patches and dark shadows of the forest produces a visual pattern very similar to the colour pattern of his back. Then, only the most alert observers might detect a Swinhoe's pheasant moving about in the interior of a forest, picking its way among clumps of vegetation, turning over litter in search of tasty morsels of food.

Swinhoe's pheasants seem to move about the forest, following the same routes day in and day out. They tend to come in and out of the forest using definite entrances among roadside vegetation. These routes can be identified by sharp-eyed observers as 'bird trails'. Researchers have seen individuals, after being startled by people, run a fairly long distance to get to an entrance.

Swinhoe's pheasants roost in trees at night like most pheasants. Researchers working in Yushan National Park have seen pheasants roosting individually, as well as witnessing a female roosting on a tree with her two large young. Radio-telemetry work found Swinhoe's pheasants roosting in the canopy of large trees. Very young chicks probably cannot jump high enough to get onto a tree, so, young chicks, one week old, still return to their nest to spend the night. But by the time they are one month old, they start roosting in trees.

Swinhoe's pheasants are omnivorous. The stomach content of a male Swinhoe's shot by a hunter in 1988 turned out 80% plant food, which included flowers, fruits, mosses, grass seeds, etc. The other 20% included termites, termite eggs, moths, grasshoppers, and beetles. Many plants and animals have been recorded as their food items and there is a seasonal variation in the availability of these. Pheasants were seen pecking at the road surface and at the vegetation along the road edge, or feeding in the herbaceous ground cover and brush in secondary growth at the forest edge. When Swinhoe's pheasants forage, they tend to move along and pause momentarily to peck here and there instead of concentrating on one spot.

If a desired food item is above its normal reach, a Swinhoe's pheasant might leap up to reach it. In the forest, they usually feed by digging with their bill, seldom scratching for food like chickens. They might use their toes to pick apart the dry leaf cover only enough to reach the food, or use areas where the litter has already been opened up by hill-partridges, or bamboo partridges.

Upon perceiving danger, such as encountering people on a trail or logging road, a Swinhoe's pheasant might freeze momentarily, then walk away deliberately, quickly, or run away. It usually holds an erect posture when fleeing, fanning its tail as it runs away. It is normally silent, but when it is extremely frightened, before or during flushing, it might utter high-pitched sharp calls. When a Swinhoe's pheasant flushes down a slope, it uses flap-and-glide wing beats. When a female is with young, she may utter a sharp "tsee" call to alert her young of danger. Young pheasants usually scatter rapidly into the vegetation on both sides of the trail, while the female moves slowly forward as if to lead the predator away from the young. After a certain distance, she slows down and calls her young together, before the whole family moves away from the trail.

There have been very few direct observations of predation on Swinhoe's pheasants. A crested goshawk was seen eating a female Swinhoe's pheasant in central Taiwan on a cloudy day. A Lesser sparrow hawk was seen eating a bird on a tree. This bird had white and dark feathers very like that of a male Swinhoe's. Local people talk about brown wood owl or tawny wood owl eating Swinhoe's pheasants. Certainly these nocturnal predators are capable of catching Swinhoe's young. Snakes may prey on Swinhoe's nestling or eggs. Gem-faced civets, Chinese civets, and other mammals are reported to be their predators too.

In locations with high concentrations of food, a number of Swinhoe's pheasants may feed together. In these feeding congregations, individuals peck at each other to maintain individual spacing. Tree squirrels sometimes forage in the vicinity of Swinhoe's pheasants. Swinhoe's pheasants would also peck at tree squirrels which are feeding too close to them. Once a squirrel got too close to a young pheasant female, and the pheasant forcefully pecked at the base of the squirrel's tail, causing it to leap to a tree and remain there.

Besides occasionally feeding together, Swinhoe's pheasants usually appear singly, except when breeding. Pairs, two females, a female with young, or several sub-adults together can be found from April through October. Early in the breeding season, courtship can sometimes be seen. Four types of courtship behaviour have been described.

1.   The male leans forward and hops around the female. In the process, he droops the wing nearer the female and lifts the opposite wing, so his colourful back and shoulders are entirely exposed to the female. The red skin on his face is fully expanded, with fleshy horns erect above and lobes hanging down below his eyes. He makes "tse, tse" calls while encircling the female. The female walks around without paying overt attention to his display.

2.   A male stands near a sitting female and beats his wings. The red skin of his face is expanded, his white crown feathers erect, and his neck stiff and swollen. After wing beating, he makes "check check" calls. The female observes his display, periodically utters "un, un, un" sounds, and preens her feathers.

3.   A male walks alongside a female and slowly circles her making "pi, pi, pi" calls in the process. When he encircles the female, his head is erect, and his wings horizontal, vibrating continuously. His tail feathers are spread out like a fan, held vertically. During this process, he

periodically utters a sharp prolonged "sui" call. Sometimes he also makes a pleasant trilling sound.

4.   A male chases after a female. When the female stops, he slowly circles around her, vibrating his wings or spreading his wings, with his tail fanning up and down continuously. One can hear rapid trilling sounds periodically.

No one knows if Swinhoe's pheasants establish breeding territories. But based on observations in the wild, it seems their activity ranges overlap even in the breeding season.

A radio-tagged female used a 25 ha area when she was bringing up four young. At least one male was using this area at the same time. No one knows if he is the only other adult in the area, or whether the size of this area is sufficient for her year-round needs, and whether it is the normal size for this species.

Male and female Swinhoe's pheasants start to appear together in February. Nests are found from February through July, but most are found in April and May. Clutch sizes vary between three and eight eggs, but most are four to five eggs. Females incubate alone and may stay for days in their nests without moving. Hatching period usually lasts two to three days. Young are precocial, but a female does not leave her nest until all the eggs hatch.

The nest of Swinhoe's pheasant remained a challenge to science until April 1995, when the first active nest was discovered and documented by researchers in Yushan National Park. [Even though a number of nests have been found by chance by people, all previous efforts by researchers failed to uncover one]. The well-hidden nature of the nest, the beautiful camouflage of the female's plumage, and the secretive nature of her behavior, all contribute to make this discovery impossible without the aid of radio-telemetry. The nests themselves are usually constructed of grass on the outside, lined with some dry leaves and feathers. Swinhoe's pheasant nests are in dark and obscure locations, generally under some large shelter, such as a fallen log, rock, or overhang in hardwood or mixed forests. All known nests are in places that are well-sheltered from rain. Ground nests were found on small depressions on slopes often as steep as 50-60 degrees. Above the nest was dense undergrowth, with thorny vegetation or ferns hanging down, making the nest virtually impossible to see from the outside. Some nests are entirely covered by dense grass on steep slopes, in conifer plantations. A number of these nests were discovered when workers were cutting grass in the plantations and flushed the females. Some nests are on trees, either in a safe depression, or well hidden by epiphytes.

In captivity it takes 25-28 days for the eggs to hatch. An observer gave interesting reports about how the chicks leave their nests on trees. From a nest that was not too high up, he saw them hop down one by one after their mother. From another nest that was several meters above the ground, he saw the mother hop out of the nest with all her young tucked under her wings. Shortly before reaching the ground, she opened her wings to break her fall, and all the young fell out. But by then they were already fairly close to the ground thus they landed safely. This observation took place in an aviary and its applicability in nature needs to be evaluated.

Sub-adult pheasants can frequently be seen foraging in small flocks in the vicinity of their mothers. However, once they reach about two thirds the size of adult female, they seem to have generally gained independence. They usually move about by themselves, and are rarely seen with adult females. Adult males appearing together with juveniles of any age, is a rare sight. There is no information about the amount of time it takes a wild Swinhoe's pheasants to mature into adulthood. However, given the similarity of this species to domestic chicken, young Swinhoe's probably do join the breeding population in one year.

Hatchings are covered in tan coloured down, which is changed into juvenile feathers at about three weeks of age. They are now in the coloration of female pheasants. This plumage is retained until the following winter, when young males start moulting into the dark blue of adult males. Thus, in winter or very early spring, it is possible to see sub-adult males in mottled blue/brown plumage. But even before this stage, one can always separate young males from young females by examining their legs. Those with small spurs are the males.

Within a national park, hunting and trapping is prohibited. More mountain climbers and visitors to Yushan National Park have reported the thrills of seeing the beautiful Swinhoe's pheasant. This proves that wild animals do learn and decrease their level of wariness in places that are safe. It would be a wonderful day when Swinhoe's pheasants in Yushan National Park can accept the presence of people without any alarm. Then one would not need to be an expert bird watcher to marvel at the details of its beauty. This is an attainable goal which needs the co-operation of all the people who visit this park.

Suitable habitat for Swinhoe's pheasants has been decreasing steadily outside of protected areas. If this continues, there is a danger that someday parks and reserves will contain their only remaining habitat. In that case each park will have an isolated population with no gene flow, which can seriously influence the genetic health of the species. If we value this beautiful pheasant and want it to thrive in Taiwan, we must not think everything is well taken care of simply because some parks or reserves have been established. We should ask: Do we really need to "tame" every piece of land not currently protected by law? Have we not found pleasures in nature which are not available in any human environment?

[Reproduced from the booklet 'The Island Treasure' published by Yushan National Park Headquarters].

Lucia Liu Severinghaus,

Academia Sinica, Taipei.

5.   WILDLIFE CONSERVATION SOCIETY SURVEYS FOR BORNEAN PEACOCK PHEASANT

The Pheasants: Status Survey and Conservation Action Plan 1995-1999 identified the Bornean peacock-pheasant *Polyplectron schleiermacheri* as a Galliform species requiring urgent conservation action. In response, a joint Wildlife Conservation Society (WCS), BirdLife International, and Indonesian Department of Forest Protection and Nature Preservation (PHPA) initiative has been conducting surveys in East and Central Kalimantan (Indonesian Borneo) to determine the current status and distribution of the Bornean peacock-pheasant.

The Bornean peacock-pheasant is one of the most elusive pheasants in the world. First collected near Muarateweh, Central Kalimantan, the species was described by Briiggemann in the 1870s and has rarely been observed by biologists. There were only 8 confirmed localities on record, along with an additional 4 or 5 locations based on reliable secondhand information. While these records span the 900 km distance from Pontianak in West Kalimantan to Samarinda in East Kalimantan, most of the reports come from Central Kalimantan. Recent hunting surveys carried out by the WCS Malaysia Program in Sabah and Sarawak indicate that local Penan and other Dayak groups had no recent knowledge of Bornean peacock-pheasants at the survey sites.

Knowledge of the Bornean peacock-pheasant's behaviour and habits is extremely limited, and no systematic studies of the species are available. It is believed to be active at dawn, to feed on fruits, seeds, and invertebrates, and to make seasonal movements in response to fruit availability. Bornean peacock-pheasants are believed to inhabit lowland rainforest, with a preference for alluvial lowland forest. This forest type is an extremely threatened habitat on Borneo because it is favoured by logging companies and shifting cultivators. The Bornean peacock-pheasant is considered rare or absent from peat and swamp forests that comprise much of the lowland forests of Borneo, and although it occurs above 305 m, most records are from lower elevations. Because of the paucity of information, the conservation status of the Bornean peacock-pheasant remains uncertain. Given its low density, and presumed preference for a seriously threatened forest type, the belief that Bornean peacock-pheasants are critically endangered may be warranted.

With a donation from Richard Olsen to WCS and additional funding from WCS, we initiated surveys to assess the conservation status of Bornean peacock-pheasants, in order to develop recommendations for *in situ* conservation and possible establishment of a captive breeding programme. In 1995, WCS awarded a grant to Mr. Resit Sozer to survey Bornean peacock pheasants along the Mahakam River, the largest river system in East Kalimantan. Mr. Sozer visited 29 sites along the Mahakam River in east Kalimantan between December 1995 and March 1996, conducting interviews and collecting information on a number of galliform species, especially Bornean peacock-pheasant. Mr. Sozer reported hearing calls resembling the closely related Malayan peacock-pheasant at two locations. At one of these locations, feathers also were collected. Although his final report is not complete, our conclusion from Mr. Sozer's preliminary report is that the Bornean peacock-pheasant is extremely rare on the Mahakam River. Further surveys are needed in other major river systems, especially the Kayan River drainage to establish the bird's status in East Kalimantan.

In Central Kalimantan, a joint WCS-BirdLife team attempted to determine the conservation status of the Bornean peacock-pheasant using a 'rapid rural appraisal' approach. A survey team composed of Nurul Laksmi Winarni (WCS), Andi Setiadi (WCS), and Iwan Setiawan (BirdLife) distributed illustrated questionnaires, and conducted standardised semi-structured interviews during July-August and November-December 1996, to find out what the local people knew about the pheasant. The questionnaire served as an educational tool as well as a research tool. The illustration of the pheasant carried a conservation message and could be removed from the questionnaire and kept by the respondent. The teams surveyed six of the seven major river drainages in Central Kalimantan, distributing nearly 1000 questionnaires and conducting more than 150 interviews in 93 villages. We received responses to 842 of the questionnaires.

Preliminary results confirm some of the speculation regarding habitat preference and distribution. Generally, most of the people in the upper reaches of the rivers were familiar with the Bornean peacock-pheasant whereas people along the lower reaches of the rivers did not recognise the bird. This appears to confirm earlier observations regarding absence from swamp forest habitats. Reports of Bornean peacock-pheasants were confirmed by collection of feathers at four locations. These areas are characterised as undisturbed forests on rich soil. In addition, we received reliable reports that the Bornean peacock pheasant had been trapped within the past 12 months at an additional 28 locations. Almost all information results from trapping efforts by local people who report that the bird was snared for meat. These initial results suggest that the Bornean peacock-pheasant is more widespread than previously thought. Most records are from 100 to 500 m elevation. The distribution of Bornean peacock-pheasant appears to increase from east to west along the foothills of the Schwaner and Muller mountain ranges. We also confirmed the presence of Bornean peacock-pheasants in the Bukit Baka-Bukit Raya National Park. These observations are significant in that they are the first reports of the bird's occurrence in a protected area.

We plan to continue our surveys during the next 12 months, pending funding. We are preparing a GIS database to plot pheasant locations and overlay forest type, soil type, elevation, presence of logging concessions, and other information to better assess the habitat preference and conservation status, and to identify priority areas for future work. This information may be used to determine if the Bornean peacock-pheasant is adequately protected in the existing protected areas of Kalimantan or if additional areas are required.

Timothy G. O'Brien and Margaret F. Kinnaird

Wildlife Cons. Soc. – Indonesia Program P. O. Box 311

Jl. Ciremei No. 8, Bogor 16003, Indonesia.

6. SATYR TRAGOPAN IN THE SINGHALILA NATIONAL PARK, DARJEELING, INDIA

Lying at the north-western extremity of the Darjeeling district of West Bengal are the beautiful oak-rhododendron-fir forests of the Singhalila National Park, (SNP) - home to the attractive yet little studied Satyr tragopan *Tragopan satyra* and the red panda *Ailurus fulgens*. Satyr tragopan is a "Vulnerable" species, and has the widest range distribution among the tragopans in India. It extends from Garhwal through Kumaon, Nepal, Sikkim, Darjeeling hills, and Bhutan, to western Arunachal Pradesh at altitudes between 2,400 m and 4,300 m. In Darjeeling hills, confirmed reports are available from the Singhalila NP, while reports from 2 other Protected Areas of the district are not confirmed. The Satyr tragopan inhabits steep, densely forested slopes with a well developed under storey of ringal bamboo *Arundinaria* sp. They are very shy and wary, and rapidly retreat for cover when encountered in the forests. Most of the time I found them solitary or in pairs and, on one or two occasions a female with chicks.

The 108.77 km² Singhalila NP derives its name from the Singhalila spur which arises abruptly from the terai plains and runs 100 km from south to north separating Darjeeling district from Sikkim and Nepal. The spur culminates at its northern extremity in some of the loftiest peaks of the world, including Kanchandzonga (8586 m), and others which are above 6000 m in height. Wet and moist temperate, and alpine forests form the main forest cover of the Singhalila NP. Climatically, the Singhalila NP is considered to be a temperate to subalpine zone. Moist conditions prevail throughout the year. The Singhalila NP has a rich diversity of flora and fauna, including birds. Most of this rich biodiversity of the Park is yet to be scientifically documented.

The main objectives of my studies on the Satyr were, to identify the pockets of distribution, and see how abundant the species was in the National Park (NP). I chose an intensive study area for detailed studies on other aspects of its ecology and behaviour, such as habitat use, feeding ecology, calling, and social behaviour. The Satyr is a very elusive bird and in the ensuing months of study, I had to depend a lot on indirect evidence of the bird (droppings, feathers, and calls) along with the direct sightings. The forests at the southern and southeastern portion of the NP were found to be the main distribution areas of the Satyr tragopan, and it was in these sites that the intensive studies were carried out. The wet temperate forests of the NP comprise the main habitat of the Satyr at altitudes between 2600-3100 m. This includes *Quercus pachyphylla*, *Rhododendron* spp., *Magnolia campbelli*, *Sorbus cuspidate*, and *Acer* spp. as the dominant tree species. Ringal bamboo (*Arundinaria maling* and *Thammocalamus aristata*) forms the dominant understory species along with small sized trees of *Rhododendron* spp., *Osmanthus* sp., *Viburnum* sp., and *Symplocos* sp. Dry mixed forests, alpine forests, pure rhododendron forests, and some plantation forest were also used by the bird.

Call counts were conducted in the intensive study area during the spring of 1995 and 1996. The male birds began calling by the end of March, and continued until May, with the peak of calling in April. 28 birds were counted in the spring of 1995 and 18 in 1996. The dawn chorus began by 0430 hrs, and it was completed by 0600 hrs, with the birds calling from their roosts, and moving around to guard their territory. Because of the difficult terrain and the very shy behavior of the species, studies on other aspects of its breeding behavior could not be carried out in detail.

In spring this year there is a programme to study the Satyrs by radio telemetry which, I hope, will provide information on many aspects of behavior and ecology of the species which is as yet unknown to us.

In order to study the feeding ecology of the Satyr tragopan, faecal droppings have been collected and preserved for subsequent faecal analysis. I personally observed the bird feeding on *Theropogon* sp., *Pilea* sp., and a fern, and on one occasion we found the seed of *Symplocos* sp. fruit in one of the droppings. Records of the phenophases of different plant species that may turn out to be the species' food have been kept on a monthly basis.

Fig. 5.2 A. A male Satyr tragopan in a breeding center for Galliformes, in Braga, Portugal. Photo by Hugo Barbosa. B. A female Satyr tragopan resting on the ground in Artis Zoo, Netherlands, 2008. Photo by Arjan Haverkamp.

Satyr tragopan males were mostly solitary or in pairs when encountered in the forest, and sometimes 2 females were observed with a single male. Females with chicks were also sighted during the studies but this was very rare.

It would be worthwhile here to mention the threats to the Singhalila NP, which in turn are threats to the Satyr. The western border of the

Singhalila NP coincides with the international border with Nepal, and is totally depleted of forest cover due to the establishment of settlements and rampant tree-felling for agriculture. There is not even a single patch of contiguous forest between Singhalila NP and Nepal. Along the 42 km border, there are 8 Nepalese settlements on the border and many other villages barely 2-3 km away from the NP. On the Indian side, the settlements are quite far from the core area of the NP, but the forests on this side are also being depleted due to new settlements and agriculture. Since the forests of the Singhalila NP did not have Protected Area status before 1992, they were under tremendous pressures in the form of devastating forest fires, large logging operations, major construction works, illegal felling and sawing, intense grazing, and large scale poaching. Habitat destruction and poaching are the two main threats to the Satyr population in the Singhalila NP. All the settlements along the border of the NP have a number of livestock, and there are yak stations (pastoral herd stations) which use the NP forests as their grazing grounds. Settlements totally depend on the NP forests to meet their fuel wood needs. Collection of forest produce, such as bamboo shoots, medicinal herbs, and bamboo stems for construction work is quite intense.

The Singhalila NP has always been a popular trekking area, and there is a large inflow of tourists, both foreign and domestic, which indirectly increases the use of the forests resources. These activities all contribute to the loss of good Satyr tragopan habitat in the NP. Poaching of Satyr tragopan was quite common before the Singhalila NP forests were given Protected Area status. Poaching was mainly carried out for recreation, for food and for supplying animals to the local zoo. Poaching activities in the NP cannot be ruled out even now. These pheasants are very easy to snare and poachers know exactly where the pheasants occur. Regular monitoring of vulnerable sites, control of yak grazing and tourism, vigilant control of poaching activities, and an overall awareness campaign, would help to lessen the intensity of the threats to the bird. No other sites of Satyr distribution have been explored in the Darjeeling hills, although there are reports of the bird elsewhere. Future surveys will perhaps be conducted in these so far unexplored areas of the Satyr distributional range.

This West Bengal Forest Department (Wildlife Circle) sponsored Project is being assisted by the World Pheasant Association.

Sarala Khaling

Centre of Wildlife and Ornithology

Aligarh Muslim University

Aligarh – 202002, India.

7. Status of the Grey jjunglefowl *Gallus sonneratii* in Periyar Tiger Reserve, Kerala, South India

The grey junglefowl *Gallus sonneratii*, an endemic species of peninsular India, was very common in Kerala in the earlier part of the 19th century. It enjoyed a wide distribution in the foothills and hills of Kerala, being more common in the foothills. The occurrence of a tiny population in northern Kerala, near the Cannanore coast, indicates that in the past, the species had been widespread in Kerala from the coast to the high hills. Today it has disappeared from the coastal plains and a vast area of the foothills and hills. Although the species is now found in forests and plantations, a fairly good number only survive in Protected Areas, mostly in the hills. The junglefowl's decline is due to hunting for food and habitat destruction. Its persecution must have started in the 1870s. Today the grey junglefowl is trapped and its eggs are regularly collected, even from Protected Areas. Since the local concept of wildlife management is mostly centered on large mammals, species like the grey junglefowl have received little management attention. There are no incidences of any person being charged for killing the species or for collecting its eggs.

Periyar Tiger Reserve is situated in the Western Ghats in Kerala between 9° 16' and 9°40'N and between 76⁰55'E and 77°25'E. The reserve has an undulating terrain varying between 200 and 1800 m, with a few peaks rising above 2000 m. Periyar has a humid climate, with temperatures ranging from 15⁰C to 31°C, and an average rainfall of 2500 mm. July has the heaviest rainfall. November through January is the cooler time of the year.

Fig 5.3 A male grey junglefowl from *A Hand-book to the Game-birds* Vol. I and II, by W.R. Ogilvie-Grant, 1897.

The following types of vegetation have been identified in the reserve: tropical evergreen forests (305 km²), tropical semi-evergreen forests (275 km²), moist deciduous forests (100 km²), grasslands (12 km²), eucalyptus plantations (55 km²), and reeds (5km²). Savannah occurs in several places. The high hills are covered with grasses.

A study of the status, habitat requirements, and population density of the grey junglefowl was started in October 1966, with a view to identifying the species' management problems, creating awareness of the need for the bird's conservation, and bringing these matters to the attention of the forest department.

The species has a wide distribution in Periyar in different forest types, plantations, and around human habitation, being more common in mid-altitude moist deciduous forests, where there is a fairly good under storey. However it is absent or rare in Periyar's high hills, such as Mangaladevi and Kumari Kulam.

The grey junglefowl is a shy bird. At the slightest disturbance it takes shelter under cover. It is often even difficult to observe from a distance. The birds mainly forage in the mornings and around noon, with activity slackening in the late afternoon. They scratch the ground with their feet, and dig with the beak, also often in the dung of elephant, gaur, and cattle, and at the base of larger trees, and areas dug by wild boar. They seem to have a preference for herbivore dung. Preliminary observations on their faecal samples have shown that grey junglefowl consume a number of insects, mostly beetles, and also seeds and grain. Crop contents of a dead female included cockroaches, weevils, ants, termites, Chrysomelid beetles, scavenger beetles, maggots, and seeds of *Psidium* spp., *Cyrtococcum oxyphyllum*, *Lantana camara,* and *Panicum repens.* Grey junglefowl also eat rice and other kitchen scraps from around residential quarters.

In Periyar, the junglefowl was observed feeding with barking deer and red spurfowl *Galloperdix spadicea.* Fighting was occasionally observed between a female junglefowl and a male spurfowl for unknown reasons. Dust-bathing and preening were observed on a few occasions, mostly around midday. The junglefowl roosted on small trees, such as Acacia, often encircled by climbers and at a height of about 20-25 ft. On two occasions, an adult male was seen perched on top whilst roosting. When going to roost, the birds fly to the lower branches and then move up to the upper branches by jumping.

The cocks were heard calling almost throughout the year, but less frequently during the monsoon, and when the birds are nesting. No information is available about the junglefowl's social organisation. Males are found alone, either with a single female or often with more than one female. This is an indication of polygamy. The adult male:female sex ratio in Periyar, however, seems almost 1:1.5. On several occasions, all-male groups comprising four to six adult males were found, mostly during June - after the birds' breeding season. Very little information is available on the species' breeding behaviour. No courtship behaviour was observed. Fighting between adult males was noted on several occasions in December, occasionally in the presence of a female. They breed between February and July in Kerala. In Periyar, we have observed junglefowl eggs from November to May. Clutch size varied from 2 to 5. Eggs are mainly laid at the base of trees and along the side of trails.

No instances of predation on the junglefowl were observed in Periyar, although domestic cats and stray dogs were seen chasing birds. On a few occasions, clusters of the bird's feathers were observed on the ground. Feathers of the junglefowl were identified in the scats of jungle cats.

Poaching of adults, and egg collection, seem to be the grey junglefowl's major management problems in Periyar. Local people, and a nomadic tribe (Malankuravar), are known to trap junglefowl from border areas of the reserve, using various types of traps. Firewood collectors often take eggs, and indiscriminately remove the under storey vegetation that is essential for the species' survival. The disturbance of incubating birds by an increasing number of domestic cattle is another threat in the reserve's border areas.

V.J. Zacharias,

Dept. of Zoology, St. Josephs College,

Devagiri, Calicut, Kerala 673008

India

8.   Hume's pheasant sightings in Mizoram, India

6 species of pheasants are known from Mizoram - Blyth's tragopan *Tragopan blythii*, red junglefowl *Gallus gallus*, kalij pheasant *Lophura leucomelana*, Hume's pheasant *Syrmaticus humiae*, and grey peacock pheasant *Polypectron bicalcaratum*. Hume's pheasant is classified as Endangered, and is included in Schedule I of the Wildlife (Protection) Act of India. In India, it has been reported from Manipur, Patkai, Naga, and Mizo (Lushai) hills.

During surveys conducted by WPA-South Asia Regional Office, Hume's pheasant has been encountered in Murlen National Park, in eastern Mizoram. Another well protected area in Mizoram is the Blue Mountain National Park, located in the southeastern part of the state at N 22° 39' E 93°02'. Hume's pheasant (known locally as 'vavu') has been reported from this region in 1950s. Though not recorded in later surveys, the local villagers and forest Department staff insisted that it still occurred in the national park.

During my 8 month stay in the area, I sighted Hume's pheasant twice. Both sightings were in dense forest. The species' habitat is mixed open forest interspersed with grass and bracken patches. Both sightings were in forests on steep hillside, and undergrowth in both areas was moderate. In recent years, Hume's has been reported from the Chin Hills, where locals thought it more common than Blyth's tragopan. However, in Blue Mountain NP, Hume's appears to be the rarer of the 2 species. Moreover, while Hume's was located in dense forest only, Blyth's tragopan was found to occur in both thick primary forest and secondary forest, as well as on steep hillsides.

Dipankar Ghose, Asst. Project Officer, WWF-India,

West Bengal State Office, Calcutta

9.   Surveys for Sclater's Monal in Northerwestern Yunnan

Introduction

We conducted surveys for Sclater's monal *Lophophura sclateri* at 3 sites along China's mountainous southwestern border with Myanmar, near the eastern range limit of Sclater's monal. Much of the survey area had only recently been opened to foreign travelers. Potential monal habitats were widespread throughout the area, but the species had been confirmed at only a handful of locations, primarily because access has been very limited.

Objectives

Our objectives were three-fold: 1) to assess variation in Sclater's monal habitats along a north-south gradient through western Yunnan, 2) to compare densities of monal under different habitat conditions, and 3) to identify a potential location for in-depth ecological studies of the species. We were joined by staff members of the Mammalogy Section of the Kunming Institute of Zoology, our host institution in China. Our work was supported by the Zoological Society of San Diego and the Kunming Institute of Zoology.

Results

We began fieldwork in early May, 1999, at Da Yang Tian, Tengchong County, near the southern limit of the monal's known distribution. Da Yang Tian is a mosaic of meadows and bamboo thickets at about 3,900 meters elevation. It is probably the most extensive patch of alpine habitat in the southern Gaoligong Range. We surveyed Da Yang Tian's meadows for 9 days, working from a series of bivouacs along the crest of the main ridge. In essence we conducted a line transect survey in slow motion. Every second day or so we relocated our camp several hundred yards further along the ridge. Before first light, we entered makeshift blinds positioned to provide a wide view of local terrain. Weather permitting, we would spend 4 hours each morning and afternoon watching and listening for monal from these blinds.

We observed between 3 and 5 adult male monals in the Da Yang Tian area. We weren't certain of the exact number, however, because we were unable to distinguish between individual males on successive days. We surmise from several hours of behavioral observation that territoriality among male monals had waned by early May. Individual males often flew cross-slope several hundred yards to alight where we had seen or heard other males. Vocalizations by newly arrived males did not elicit a vocal response, however, from the males already present. On the 3rd day of fieldwork, an unseasonably early monsoon storm struck. We held out through 5 days of persistent rain and fog, but found it impossible to observe additional monals.

On May 23rd we arrived at Ci Kai, Gongshan County, the market town nearest our northernmost survey site, Dong Shao Fang. Chinese scientists had collected monal at Dong Shao Fang in years past, and a well-established trail provided relatively easy access to alpine elevations. Our departure into the field was delayed, however, by continued heavy rains, which raised the Pula River above local footpaths. Once we arrived at Dong Shao Fang, rain and fog persisted for 5 more days, largely restricting us to our tents and kitchen. It was becoming clear that a springtime study of Sclater's monal was going to require special monsoon-worthy equipment, as well as an innovative approach to collecting data. On May 30th, the weather subsided enough for us to check a nearby 4,000 metre pass for signs of monal. Two members of our party reported hearing the call of a male monal, but the rain intensified and we were forced to return to camp before others could confirm its presence. At this site too, inclement weather essentially confined us to our tents on most of the days allocated for a survey.

In early June, we arrived at our final survey site, on the slopes above a dilapidated lumber town named Zhiziluo, central Fugong County. Chinese scientists had collected many Sclater's monal here over the years. It was our central survey site, approximately midway between Da Yang Tian and Dong Shao Fang. We endured more foul weather on the local summit for a full 3 days. On the 4th day, we descended to the upper tree limit, where we had observed what we believed were monal fecal droppings. On the way, we virtually stumbled into a female monal with 3 two-week-old chicks. Over the next day and a half, we searched nearby slopes for more signs of monal. At 4,100 metres elevation on a prominent spur, we noticed what looked like an unusual kind of soil erosion or patterned ground. On closer inspection, we saw that the soil's mossy crust had been turned over in 1 to 2 inch squares. Fresh monal droppings and deeply-excavated pits confirmed that we had found a site where a large number of monal had recently fed. The diggings extended for several hundred meters up the spur. We were delighted to find examples of a fungus-infected caterpillar within the pits dug by monal. This 'dong chong xia cao' is a popular and expensive traditional Chinese medication, that David Rimlinger has previously determined to be a likely food item of the rare Chinese monal in Sichuan. Other 'table scraps' left behind by the monal included stems and roots of cinquefoil (Rosaceae), buttercup (Ranunculaceae), jack-in-the-pulpit (Araceae), and fritillary (Liliaceae). Other evidence suggested that the monal had also eaten grass shoots, a plant of the carrot family, beetle larvae, and the larvae of wood-boring insects. The proximity of the diggings to dense thickets, both here and at Da Yang Tian, suggest that Sclater's Monal often forage within a few meters of bamboo or broadleaf thickets. The presumed flock of monal remained quiet and out of sight throughout our stay. We suspect their vocal season had passed by the time we arrived at Zhiziluo in early June.

We decided the Zhiziluo site was the best of the 3 sites for an in-depth monal study. It offered a diverse mosaic of habitat types, a branching ridge system that would facilitate radio-telemetry, expert and congenial guides, and an apparent abundance of monal. Da Yang Tian was appealing for its relative ease of access, abundance of interesting wildlife including- red panda, Himalayan black bear, takin, tufted deer, and blood pheasants. Local authorities were hospitable, and the local field help was pretty good under the circumstances. But the extent of monal habitat and the numbers of monals at Da Yang Tian appeared to be quite limited. Dong Shao Fang, our northernmost site, had such good access that it was probably too disturbed by humans. The phenomenal rainfall at Dong Shao Fang would be a formidable obstacle to overcome, and the local help we encountered considered monal research little more than a business opportunity.

Discussion

So far as we could determine, Sclater's monal persist throughout their historic range in Yunnan. We also received reports that the species occurs even further south than currently recognized. Han Lianxian of Yunnan's Southwest Forestry College is currently checking several potential range extensions. The species' occurrence on discontinuous mountain summits, however, suggests it could be vulnerable to local extinction. This is particularly true in the south, where alpine habitats occur as isolated islands, and local populations may consist of as few as a dozen individuals.

Our surveys produced 3 to 5 adult males at Da Yang Tian, one vocalizing individual at Dong Shao Fang, and one female with 3 chicks at Zhiziluo. Unfortunately, bad weather and our failure to fully implement the intended study design, leave us with no means to compare the abundance of monal at the 3 sites we visited. All the birds we encountered occupied large meadows or meadow complexes that extended several hundred metres down steep slopes. These meadows were surrounded by bamboo or rhododendron thickets, with at least a few rock outcrops present. The lowest elevation at which we observed such meadows, as well as monal, was about 3,000 metres. The flowering meadow plants we observed at all three sites included cinquefoil, buttercup, fritillary, jack-in-the-pulpit, and peony (Rosaceae). Many bulb-forming monocots occurred within the bamboo and rhododendron thickets as well. Certain habitat features increased or decreased along the north-south axis of the monal's Yunnan distribution. Alpine habitats tend to be more contiguous and rockier at more northerly locations because mountain ranges of the region achieve greater elevations further north. Rainfall is also higher further north. Alpine thickets at more northerly locations contained a greater proportion of broadleaf shrubs and less bamboo.

While travelling between our 3 survey sites, we encountered 2 dead monals for sale in outdoor markets. Even though the species receives the same protected status as giant pandas, carrying a potential death sentence, monal were being sold openly for about $12. It appears that illegal market hunting still poses a significant threat to Sclater's monal. Herb collection for traditional Chinese medicine may also be detrimental to monal, but not because herb collectors compete with monal for the roots and tubers of such plants as "ta huang" (wild rhubarb) or "bei mu" (fritillary). Herb collectors are often the only humans that visit the heights occupied by Sclater's monal. They camp for several days while collecting herbs, and many set trap-lines or hunt with guns to add meat to their meals. China's burgeoning rural economy is enticing increasing numbers of entrepreneurs into the mountains to harvest such 'alternative forest products' as medicinal herbs. The apparent connection between these cottage industries and the illegal harvest of large animals like Sclater's monal deserves serious consideration by China's wildlife conservation authorities.

David Rimlinger, James Bland, Wen Xianji, Yang Xiaojun.

10.   Status and habitat of Cheer pheasant in Himachal Pradesh, India: Results of 1997-1998 surveys.

Introduction

The habitat of cheer pheasant *Catreus wallichii* is distributed in India, Pakistan, and Nepal. In India, it has been found at numerous sites in Himachal Pradesh and Uttar Pradesh (now in Uttranchal). It inhabits steep hillsides with scrub and grass, dissected with wooded ravines at 1200-3500 m, and has a strong affinity for early successional habitats maintained by frequent human intervention. Cheer pheasant populations have been severely reduced due to habitat degradation, over-hunting, and conversion of land for agriculture. Due to its specialised habitat requirements, the distribution of cheer is very patchy. Most known populations are very small (<10 birds), making them extremely vulnerable to local extinction.

The objective of these surveys was to document the status of cheer in previously known sites, and to find new sites in Himachal Pradesh. The structure and composition of the cheer habitat were recorded during the surveys. These habitat parameters were linked to an index of population density derived from cheer call counts. Human landuse practices at sites studied were also documented.

Study areas

I conducted surveys from March to June in 1997 and 1998, in the state of Himachal Pradesh, India which lies between latitudes 30° 22' N to 33° 12' N and longitudes 75° 45' E to 79° 04' E. I selected 7 study areas in the districts of Solan, Shimla, Chamba, and Kinnaur, that collectively represent the scattered habitat of cheer in the state.

Chail Wildlife Sanctuary is an area of 108.5 km² located about 20 km south of Shimla. The sanctuary has Chail town and 121 villages within its boundaries. The area comprises Himalayan subtropical pine forest with extensive south- and west- facing grassy slopes supporting scattered Chir pine *Pinus roxburghii*, and extensive north facing slopes of Ban oak *Quercus leucotrichophora* forests. *Rhododendron arboreum*, deodar *Cedrus deodara*, Kainth *Pyrus pashia,* and blue pine *Pinus wallichiana* are also present. The undergrowth is mainly *Berberis* and *Rubus*, with some *Rosa, Daphnae, Myrsine*, and *Rhabdosia*. The habitat supported extensive patches of tall grass at places, and is regularly subject to grazing, cutting, and burning.

Majathal Wildlife Sanctuary (70 km²) lies on the south bank of the river Sutlej. The sanctuary supports typical West Himalayan low altitude forest comprising steep slopes covered with vast expanses of tall grass sparsely forested with Chir pine, Ban oak and some rhododendron. *Rubus* dominated in undergrowth of *Hypericum, Rosa, Daphnae, Berberis, Indigofera, Rhabdosia*, and *Elsholtzia*. There were some patches of *Euphorbia* within grassy tracts. Many areas were grazed and cut in Majathal, with some signs of burning in Kangri and Surgadwari West.

Kaksthal lies on the south bank of the river Sutlej in district Kinnaur, and is located 15 km from Nichar above the Shimla - Kinnaur national highway. This site probably represents the present northern limit for cheer in Himachal Pradesh. There are sparsely forested patches of blue pine, edible pine *Pinus gerardiana*, Kharsu oak *Quercus semecarpifolia*, and spruce *Picea smithiana*. Some NE slopes supported thick deodar forests. Steep SW slopes were covered with short grass and scrub comprising mainly *Rhabdosia, Rubus* and *Berberis*. A large grassy area on the SW slope is being planted with Chir pine. Kaksthal was the most disturbed of all the sites studied. There were signs of cutting, burning, and grazing, and some slopes, being denuded of cover, were heavily eroded.

Tundah Wildlife Sanctuary is an area of 64.2 km² together with which I surveyed Sara Reserve Forest and Bhaatal in district Chamba, and the Thathana Reserve Forest to the southeast. All the study areas in Chamba comprised mixed conifer forests and moist temperate deciduous forests. In Tundah Wildlife Sanctuary, the tree cover included blue pine, deodar, spruce, Kharsu oak, Banni oak *Quercus glauca*, Moru oak *Quercus floribunda*, silver fir *Abies pindrow*, walnut *Juglans regia*, horse chestnut *Aesculus indica*, and Himalayan elm *Ulmus wallichiana*. *Indigofera* and *Berberis* along with some *Rubus, Rosa, Daphnae, Cotoneaster, Lonicera*, and *Rhus* dominated the scrub cover. The Thathana Reserve Forest supported silver fir, spruce, and blue pine in tree cover and undergrowth of *Berberis, Indigofera* and *Myrsine*. In Sara Reserve Forest and Bhaatal, there were mixed forests of silver fir, spruce, and Kharsu oak, and a shrub cover of *Berberis, Rumex*, and *Rosa*. There were extensive patches of short grass with signs of cutting and burning at numerous places. Gujjar (nomadic herdsmen) settlements were found throughout all the study areas in Chamba.

Methods

Information on the cheer locations in the selected study areas was collected from published literature and through interviews with local villagers and shepherds. Dawn and dusk calls, and the call playback method, were used to estimate the number of sites where cheer called (henceforth called calling sites).

For dawn call counts, the transects were manned 30 minutes before and until 60 minutes after sunrise, and each observer noted the number of calling sites and their positions. Times of calls and compass bearings of calling sites were also used to collate the data and estimate a minimum number of calling sites per transect. The call playback method was used only when no cheer calls were heard during dawn and dusk watches at a particular site. I used only two call broadcasts of short duration (20 seconds each) to avoid the risk of disturbance to calling birds. In case of a response after a call was broadcast, the number of calling sites was noted. For estimating density indices, each calling site detected was treated as an individual data point. Only the data from dawn call counts was used, as birds appear generally less likely to call at dusk. The area sampled at each site was estimated from the length of the transect and an effective strip width of 200 m.

Habitat sampling plots were randomly located in patches of structurally different vegetation both at the calling sites and the vacant sites. Sites where no cheer calls were heard even after call playback were classed as vacant sites. The selected strata were grassland, grassland interspersed with some trees, mixed forest, conifer forest, and agricultural land. Naturally occurring trails in selected strata in cheer and vacant sites were used as transects, and sampling plots were located by generating random numbers.

At locations generated by random numbers in calling and vacant sites, 0.05 ha (radius=12.6 m) circular plots were marked. The following data were collected - diameter of trees and saplings at breast height. All trees within the circular plot were counted by species. Canopy cover (%) was estimated by taking 20 + or - readings through a sighting tube (diameter = 5 cm) for the presence or absence of green leaves.

Within each circular plot, one 4x4 m quadrat was positioned randomly for sampling shrubs. Shrub cover (%) was estimated at 3 heights: 0.5 m, 1.0 m, and 1.5 m - by counting the number of covered squares (each square = 5x5 cm) of a 30x50 cm chequered board at a distance of 5 m. Data on ground cover were collected in 2 randomly located 1x1 m quadrats marked within the circular plot, and mean values were calculated. Ground cover (%) was estimated along a diagonal through each 1x1 m quadrat by taking 20 + or — readings through a sighting tube (diameter = 3 cm) held at waist height for the presence or absence of ground cover, respectively. Ground cover height was measured at four corners of each 1x1 m quadrat with a scale. At each site, altitude, aspect, slope, number of villages within 1-2 km, and presence of a water body, cliff/ravine and cultivation within 300 m, were all noted. Land use practices with respect to cutting and burning within one year, and grazing were recorded.

Results

Index of Cheer Pheasant Density

A total of 15 calling sites were detected at Majathal Wildlife Sanctuary and the density index equated to 17 sites per km² in suitable habitat. In Chail Wildlife Sanctuary, call counts on the Khariun and Blossom beats produced 24 calling sites at a density index of 5 sites per km². Density index at Kaksthal and surrounding areas was 3 sites per km². Bhaatal and Sara Reserve Forest produced 5 and 4 sites per km², respectively. Density indices from 2 sites in Tundah Wildlife Sanctuary and 3 sites in Thathana Reserve Forest were 4 and 5 calling sites per km², respectively.

Habitat Use by Cheer Pheasant

All the cheer sites surveyed were within 2 km of the nearest human habitation and within 300 m of a cliff or ravine. Water was available within 300 m only at Bhaatal, Tundah Wildlife Sanctuary and Thathana Reserve Forest. Both the sites at Tundah Wildlife Sanctuary were near cultivated land. There were differences in certain vegetation characteristics between habitat plots at calling and vacant sites. Number of trees, saplings and shrubs was lower at calling sites than at vacant sites. All calling sites had some shrub cover whereas some vacant sites lacked it completely. Height of shrub cover at calling sites showed a greater diversity than at vacant sites. There were a significantly higher percentage of shrubs (height > 0.5—1.0 m) at cheer sites. All calling sites had a significantly greater ground cover than vacant sites.

The calling sites had 0-30 % forest cover and vacant sites had 10-50 % forest cover, and the difference was significant (p <0.05). All calling sites supported a significantly (p < 0.05) greater percentage of grass cover (30-75%) than vacant sites (20-50%). There was no significant difference between percentage of bare ground at calling sites (7 %) and vacant sites (10 %).

A correlation analysis between cheer density indices at various sites and habitat variables was carried out. Density indices were positively correlated with ground cover and shrub cover and negatively correlated with tree density, canopy cover and sapling density.

Discussion

Cheer were recorded in a wide variety of forest types but showed a strong preference for open areas with dense ground cover. The grassland habitats of cheer are successional; and numerous areas recently found to hold small populations are regularly disturbed by grass cutting, cattle grazing, and stubble burning. Most of the cheer sites were located within 1-2 km of human habitation. Locals clear vast tracts of land for their cattle by cutting and burning. This is done primarily to prevent regeneration of scrub and forest, and probably has helped to maintain the scrub-grassland habitat of cheer. I did not come across any direct evidence of poaching. However, a discrete inquiry with locals revealed that there was poaching especially in winter when cheer descend closer to villages to avoid snow.

At many sites in Kaksthal, Bhaatal, Sara Reserve Forest, Tundah Wildlife Sanctuary, and Thathana Reserve Forest, cutting and burning was so extensive that no cover was left for use by the birds even in April and May, when they breed. Cheer pheasants are easily detected by their calls, and this combined with their open habitat make them exceptionally vulnerable to hunting.

Although, the species is easily detected by call, it is extremely difficult to locate the exact spot in the habitat which the bird is using. Therefore, all habitat plots within areas of calling were presumed as being used by the bird at a particular time. Habitat variables were measured in some of these plots selected randomly. Sites where no cheer calls were heard even after call playback were selected as vacant sites. I selected vacant sites within the same study areas where calling sites were located. This was done to eliminate the effect of hunting on the data, since hunting may remove individuals of a population from an area suitable for cheer pheasant. Whereas within an area of suitable habitat, it is highly unlikely that cheer are hunted out from one site and not from others. In this study, cheer inhabitated mainly open areas with few trees and saplings, dense cover of tall grass and moderate shrub cover up to 1 m tall.

Local land use practices of seasonal grass cutting, stubble burning, and cattle grazing have probably maintained the cheer habitat. However, with an ever-increasing human population pressure, these previously seasonal activities are now, to some extent, being carried out throughout the year. Habitat degradation has assumed such proportions, that many areas were bare and eroded with practically no regeneration at all. Reducing human pressure, modifying traditional methods of landuse, and enforcing stringent penalties on poaching, will probably help to prevent the march of this species towards extinction.

Rajiv S. Kalsi

Department of Zoology

M.L.N. College, Yamuna Nagar 135001

Haryana, India.

11. The status and distribution of green peafowl *Pavo muticus* in southern Mondulkiri Province, Cambodia

Introduction

Historically, green peafowl *Pavo muticus* was distributed throughout most of the wooded lowland and lower hills of Cambodia, and ancient Khmer carved it on the temples of Angkor. Unfortunately, due to hunting for food and for trade of its tail feathers, it is now extirpated from large parts of the country. Important populations were recently discovered in Mondulkiri and Preah Vihear Provinces.

Mondulkiri Province lies in the north-east of Cambodia, much of it between 200 m and 400 m elevation. It remains one of the most forested province of Cambodia. In the south of the province, a large part has been leased as a logging concession to Samling International, and much of the remainder is in Snoul Wildlife Sanctuary. Mondulkiri Province has a human population in 1998 estimated to be 32,407, one of the lowest human densities in the country (2 person /km²).

Our survey focused mainly on the Samling concession and adjacent areas of Snoul W.S., where there are many areas of good habitat, such as evergreen forest, semi-evergreen forest, deciduous dipterocarp forest, grassland, bamboo, scrub, and many wetlands - streams, lakes and ponds.

Objectives

The survey, which will be continued and expanded during 2002 dry season, aims to:

1. Identify key areas for green peafowl conservation.

2. Investigate the species' habitat use and limiting factors.

3.   Predict the distribution of green peafowl across Mondulkiri, and to estimate its population.

Methods

The survey was carried out between 23 March and 5 April 2001. The survey used green peafowl call counts as the primary tool to investigate its distribution. The calls of green peafowl are very loud and easily heard in good conditions at distances up to about 1 km. The survey was conducted in breeding season, as at this time, the males are most vocal. Outside of this period, the birds are generally silent.

Point counts were conducted at least 2 km or more apart. The call counts were usually made in early morning (05h30 to 07h30) and in the evening from (16h30 to 18h30). Only after heavy rains were green peafowl heard calling during other hours (from 10h00 to the early afternoon). Counts were made by two observers. The compass bearing and distance, time and type of call were noted for every green peafowl call heard.

At the end of each observation period, we estimated the number of calling birds present. This was based on direction and timing of the calls. For example if 2 calls were heard from different compass bearing or distances, 2 birds were judged to be present. If we heard calls from the same distance and direction we counted only 1 bird. Sightings of peafowl or their tracks were also recorded separately, even though rare.

Results

A total of 10 independent point counts were made in 3 districts of southern Mondulkiri Province - O Reang, Keo Seima and Snoul.

Peafowl were recorded from 8 of the 10 sites. The minimum number recorded during the survey was 25 and the maximum was 30. The highest numbers were recorded in deciduous dipterocarp forest and mixed deciduous forest. They were rarely recorded in evergreen forest. Their distribution also appeared to be related to the presence of permanent water and the distance from human settlement.

In total, 162 bird species were recorded during the survey, but the only other threatened Galliform to be recorded was Germain's peacock pheasant *Polyplectron germaini*. This was locally common in the logged evergreen and semi-evergreen forest, with 5 recorded (primarily heard), including 2 direct sightings.

Discussion

This was the first species-specific survey for green peafowl in Cambodia, and there were a number of problems that limited its success. The survey was begun rather late, and by that time the calling season had nearly finished (according to local reports). It was also implemented in a small area only. The survey period was short, and it rained every day, making it difficult to move between sites during fieldwork. The number of green peafowl recorded at each site is therefore not considered to be representative of the relative density at the site. Point counts could be located only where access was possible, and the distance between points was not always 2 km, due to access constraints, habitat type, and water.

However, we conclude that large, relatively undisturbed areas of deciduous dipterocarp and mixed forest, with access to permanent water, are essential for green peafowl. This habitat is also of major importance to several large mammal species, particularly wild cattle, as well as several endangered large water birds, such as adjutant storks *Leptoptilos* sp. It is therefore, a major priority for conservation. Although Mondulkiri Province still contains a large area of forest, it is heavily disturbed by resin collectors, and local people collecting other non-timber forest products. The most significant threat to the peafowl in Mondulkiri is targeted- and incidental hunting by local people, for both domestic consumption and trade.

Recommendations

More intensive and longer field surveys using the point call count methodology are planned for the dry season of 2002. They should begin earlier in the year (i.e. early February), to coincide better with the main calling season. They should focus on unsurveyed areas of suitable habitat in O Reang District, Snoul W.S., and Phnom Prich W.S., in both Mondulkiri and Kratie Provinces. The long-term conservation of green peafowl as well as other important threatened large mammals and birds, will involve the protection of large, continuous blocks of deciduous dipterocarp and mixed forest with undisturbed access to permanent water.

Tan Setha & Pich Bunnat

WCS-Cambodia Program.

12.   A report on radio tracking of western tragopan in the Great Himalayan National Park, India

The western tragopan *Tragopan melanocephalus* is one among the poorly studied threatened pheasants of the world. The global status of the population is precariously low (<5000 individuals), and it seems that they have about only 2000-3000 km^2 for survival in the entire distribution range.

Conservation initiatives have often been handicapped by inadequate scientific support on the ecology of the species. The dearth of knowledge is such that, prior to 2 independent studies just completed in India (in 1999 and 2000), the 6 months study by Islam (1985) in Pakistan was the only intensive effort to study the ecology of the species. Perhaps its elusive behaviour and low density have resulted in highly variable and poor sighting records, which limit conclusive inferences on its ecology. In order to counter these problems, and also to obtain home range estimate for western tragopan, we initiated a study in Great Himalayan National Park (GHNP), India with radio-telemetry as the primary sampling protocol.

Attempts were made to trap the birds during spring (April-June) 1999 in Tirthan valley, using two types of locally made traps viz., 'Fall net' (n = 6) and 'Leg-holdnoose' (n = 9). A total of 12 localities were chosen for trapping, and traps were placed in previously identified sites such as water holes, roost sites, and daily movement areas, and were monitored periodically. Also, on locating or hearing the bird, the traps were set at 200 m above the bird, and the bird was chased towards the net by 3 - 4 persons forming a semicircle. A total of 256 man-days (4 persons x 64 days in 3 months) were spent trapping in the altitudinal range between 2600 m and 3000 m, where we had maximum sightings of tragopans during the past 3 years of fieldwork as part of a larger project carried out by Wildlife Institute of India. Considering the number of traps and the amount of time spent each day, the total trap hours amounted to 6694 during the trapping exercise, contributed by 3924 net hours and 2767 noose hours.

On 14 May 1999, a female western tragopan was trapped in a leg-hold noose placed in a nullah within mixed broadleaf and conifer forest above Grahani thach. The bird was fitted with a radio transmitter, using a standard Biotrack necklace type collar weighing about 50g, which had the potential life span of a minimum 12 months. The bird was radio tracked (using three element Yagi antenna and Mariner 57 Biotag receiver) until November 1999, covering both summer (May - September) and autumn (October and November) seasons, after which there was no signal obtained perhaps due to transmitter failure, or the radio tagged bird had been taken by a predator. A total of 72 radiolocations representing summer (51 locations) and autumn (21 locations) seasons were obtained, and the home range was estimated based on 100% Minimum Convex Polygon (MCP) method using the ArcView software.

The home range of the female tragopan for the entire study period was estimated to be 31.6 ha, and for summer and winter, the home ranges were 20.5 ha and 4.7 ha respectively. The bird was found to move in the elevation range 2530 – 2710 m in summer, and 2440 – 2530 m in autumn. In both the seasons, the bird was using broad leaf dominated forests, with moderate level of canopy cover and shrub density. It used areas with high tree density and shrub density. The dominant shrub species in the home range area was montane bamboo (*Thamnocalamus spathiflorus*).

Despite our intensive efforts to radio tag a minimum of 6 individuals, our trap success was limited to just 1 bird. The time selection and inexperience could possibly have contributed to low trap success. It was a great challenge to trap the birds during the breeding season, when they were largely secretive and had dispersed in wide areas. Local trappers, who claimed to have trapped the birds in earlier days, were of the same view, and were unambiguously pessimistic on trap success in this season. Nonetheless, we were certainly convinced that with the experience gained, the trap success can be increased if attempted in winter, as the birds are then concentrated in lower elevation areas; and perhaps baiting can also attract the birds to the traps due to resource crunch in this season. Future study, with adequate number of radio-tagged birds, would not only help to arrive at decisive inference on the home range and habitat preference of the species, but might also reveal other interesting facts on its ecology and social behaviour. The collective empirical data obtained from recent studies, past surveys, and through telemetry study would greatly benefit conservation of the species.

Acknowledgements

We would like to thank the World Pheasant Association, U.K., and Dr. Rahul Kaul of South Asia Field Office, New Delhi, for providing transmitters and other necessary telemetry equipment for this study. Mr. Rajat Bhargav of TRAFFIC-INDIA helped us with traps and expertise. Director, WII and colleagues of FREE-GHNP project provided support in the field. My Field Staff - Himat Ram, Pritam Singh, and Sher Singh were highly enthusiastic and without their hardship, it would have been a futile exercise.

K. Ramesh, S. Sathyakumar and G.S. Rawat

Wildlife Institute of India, Chandrabani,

Dehradun, India.

13. Studies on the abundance and distribution patterns of great argus *Argusianus argus* in Bukit Barisan Selatan National Park, Sumatra, Indonesia

Studies on the abundance and distribution patterns of great argus *Argusianus argus* were undertaken during 2001, in Bukit Barisan Selatan National Park, Sumatra, Indonesia.

The studies were conducted to achieve three objectives: 1) to determine the density and abundance of the species using distance sampling line transect method, and using those estimates to calibrate camera trap data; 2) to evaluate habitat characteristics of dancing grounds, and habitat use of male great argus; and 3) to determine the home range and movement patterns of male Great argus. Research was conducted in the Wildlife Conservation Society WCS/Perlindungan Hutan dan Konservasi Alam (PHKA) Conservation and Training Research Centre in Way Canguk area (5° 39' S; 104° 24' E), Bukit Barisan Selatan National Park (BBSNP). The park is the 3rd largest protected area in Sumatra, covering an area of 3,568 km², which functions as primary watershed for southwestern Sumatra.

To estimate the density and abundance of the pheasant, I used a distance method using line transects conducted on monthly basis during 1998-2000. Counts included calling birds and birds that were actually seen. Furthermore, assessment of camera trapping in detecting the argus was carried out. Results of line transect surveys suggested that density of argus might be increasing in the park. However, there is a need for more technique refinement. I also found that camera trapping offers an alternative method to study secretive and elusive pheasants such as the great argus.

Fig 5.4 A male great argus pheasant from *A Hand-book to the Game-birds* Vol. I and II, by W.R. Ogilvie-Grant, 1897.

With data collected during November 2001, I also evaluated habitat characteristics of dancing grounds to see whether adult males select specific sites for their dancing ground placement. I used 15 active dancing grounds and 15 random sites for comparison. At each site, 9 categorical and 9 continuous variables were measured. Males appear to select areas with open understory, which give ease and flexibility to facilitate their courtship display. Occupied sites had fewer lianas and smaller leaf size of nearest trees.

In addition to these studies, radio telemetry was used to evaluate habitat use, and to determine the home range, and movement patterns of males. 7 males fitted with necklace radio transmitters (ATS model A 3960) were tracked for 3 to 5 months during July - November 2001, depending on time of capture. Home range size was analyzed using 100% Minimum

Convex Polygon, and utilization distribution was analyzed using Kernel home range. Evaluation on habitat use indicated that males placed their home range randomly within the study area. However, undisturbed forest with large trees was identified as the most important habitat for males great argus within the home range. Home ranges were varied and larger than those found in a previous study. Home range size was around 7-32 ha. However, there were no changes in home range size between months. Even though radiolocations were concentrated around dancing ground, I found that there were no patterns of movement in relation to dancing ground over months. Core areas were between 6-12% of total home range. In addition, home range size of males was not correlated with fallen food abundance.

These results suggest that habitat is the most important factor affecting great argus spatial ecology. Habitat heterogeneity may limit the area available for home ranges and undisturbed forest with large trees, which was identified as the most important habitat, remains in only small proportions of the study area. Consequently, preservation of this part of the park, as well as other similar areas should be a higher priority.

Nurul L. Winarni

D. B. Warnell School of Forest Resources,

University of Georgia, Athens, Georgia, USA.

14. Somerset Wildlife Park to help the endangered Reeves's pheasant

The Reeves's pheasant was formerly a common and widespread species, but its range is now severely reduced and fragmented. Continued loss and fragmentation of its habitat, along with direct exploitation, means that it is classified as vulnerable to extinction, with an estimated population of less than 5,000 birds.

Fig 5.5 A male Reeves's pheasant from *A Hand-book to the Game-birds* Vol. I and II, by W.R. Ogilvie-Grant, 1897.

The tail feathers of the male can grow to almost 2 meters in length and are recorded in the Guinness Book of Records as the longest natural tail feather of any bird. Unfortunately, males are sometimes killed for their tail feathers, which are used in the head-dresses of some Beijing Opera costumes. Traditionally, these feathers were also used in the headwear of soldiers in the Chinese army, as can be seen from the soldiers' costumes in the film The Last Emperor.

Feathers from a flock of male Reeves's pheasants, kept at Cricket St. Thomas Wildlife Park, will be collected from the birds after their annual moult in July. These feathers will then be sent to The World Pheasant Association in China, where they will be donated to the Beijing Opera. In this way, wild birds will be protected and will not have to be sacrificed, whilst the traditional Chinese Opera costumes can still be made and enjoyed. It is hoped that this initiative will both reduce the illegal hunting of wild birds, and raise public awareness of the need for conservation of the birds' remaining habitat.

Long term studies of the species have been undertaken by Prof. Zheng Guang-mei and Prof. Zhang Zheng-wang at Beijing Normal University. Professor Zhang Zheng-wang is currently running a 4 year project for the Chinese Government to study the conservation requirements of Reeves's Pheasant. He and Professor Zheng Guang-mei visited the Cricket St. Thomas Wildlife Park on January 14th 2002 to open formally the Reeves's pheasant conservation area.

We need to raise funds to help improve the management of the few Protected Areas where this pheasant is found. The species would make an ideal flagship for raising public awareness of conservation in these areas, with the aim of reducing illegal hunting and egg collection.

John Corder

Somerset, UK.

Chapter 6

Conservation of Pheasants in India

Conservation projects have helped some pheasant species recover their natural populations. The latest population census of the western tragopan *Tragopan melanocephalus* in the Great Himalayan National Park in Himachal Pradesh, in year 2010, has shown an improvement in their numbers. Other species, like the Southeast Asian green peafowl *Pavo muticus*, have not been as lucky. Hunting and habitat fragmentation have caused this species' numbers to reduce further, changing its status from 'Vulnerable' to 'Endangered' in the Red List updated in year 2009.

Conservation scientists have proven that the protection of wildlife habitat accompanied by scientifically designed and managed Conservation Breeding Centers are the way ahead for saving endangered species. Initiatives in this field in Himachal Pradesh include the rationalization of Protected Areas to reduce man-animal conflicts, a plan for the Long-Term Ecological Monitoring of the GHNP, and setting up of Conservation Breeding Centers. They are supported by awareness and livelihood projects for people living near Protected Areas.

Fig. 6 A painting of a pair of Mrs. Hume's pheasants by J.E. Lodge from 'A Monograph of the Pheasants' by William Beebe, 1918-22.

Chapter 6

Conservation of Pheasants in India

Galliformes, commonly called 'game birds' due to their popularity in the sport of hunting, are represented globally by 286 species, and comprise pheasants, partridges, quails, francolins, grouse, cracids, and megapodes. India has 50 Galliformes species, 17 of which are pheasants (32 partridges, quails and francolins, and 1 megapode) (Kaul, 2007). The megapodes are a unique group of birds as they utilise external sources of heat to incubate their eggs. The family Megapodiidae consists of 22 species in seven genera, most of which are island forms and are mainly found in the Indo-Australian region east of Wallace's line. In India, the Nicobar megapode *Megapodius nicobariensis*, occurs in the Nicobar group of islands. India has a rich diversity of partridges, quails, francolins, and snowcocks. Of the species found here, the common quail *Coturnix coturnix* is migratory. Two genera contain the largest number of species in India, *Arborophila*, or the hill-partridges, and *Francolinus*, the francolins. The Himalayan quail *Ophrysia superciliosa*, is presently assessed as 'Critically Endangered', and has not been officially reported since the 1890s, but unconfirmed reports of its sightings have kept it from being declared officially 'Extinct'.

The Himalayan region of India is rich in pheasants; 14 out of the 17 Indian pheasant species are restricted to the Himalayan states. The Eastern Himalayas have 11 species of pheasants out of which 7 are exclusively found in this region. The Western Himalayas have 8 species, out of which 3 (western tragopan *Tragopan melanocephalus*, cheer *Catreus wallichi*, and koklass *Pucrasia macrolopha*) are endemic to this region. Satyr tragopan *Tragopan satyra*, Himalayan monal *Lophophorus impejanus*, kalij *Lophura leucomelanos*, and red junglefowl *Gallus gallus* are common to the two Himalayan regions. The blue peafowl *Pavo cristatus* and the red junglefowl are also found in central and peninsular India (Kaul, 2007). 6 species of Indian pheasants are listed in the IUCN 'threatened' categories (see Appendix 1).

Table 6.1: Distribution of the 17 pheasant species in India.

Region of India	Pheasant Species
Western Himalayas- J&K, HP, Utknd	8 species - western tragopan, Satyr tragopan, Himalayan monal, koklass, cheer, kalij, blue peafowl, red junglefowl
Eastern Himalayas – 9 states	12 species - Satyr tagopan, Blyth's tragopan, Temminck's tragopan, Himalayan monal, Sclater's monal, koklass, Tibetan eared pheasant, grey peacock pheasant, green peafowl (unconfirmed), blue peafowl, kalij, red junglefowl
Plains	3 species - blue peafowl, red junglefowl, grey junglefowl

The 17 species of Indian pheasants, which constitute one-third of the global total of 50 pheasant species, are among the most charismatic birds of the world. The colourful Indian peafowl *Pavo muticus* is India's National Bird; it is also the State Bird of Orissa. In fact, several States have designated different pheasant species as state birds. The Himalayan monal pheasant *Lophophorus impejanus*, "the bird of nine colours", is the state bird for Uttarakhand, and also the national bird of Nepal. While Himachal Pradesh has the western tragopan *Tragopan melanocephalus* as its State Bird, Nagaland has the Blyth's tragopan *Tragopan blythii*, Sikkim has the blood pheasant *Ithaginis cruentus*, and both Manipur and Mizoram have the Hume's pheasant *Syrmaticus humiae* as their state birds. This makes a total of 7 Indian states having different pheasant species as their state birds. Despite that, the status of several pheasant species in the country is a matter of concern. 5 pheasant species are in the 'vulnerable' category - cheer pheasant, western tragopan, Blyth's tragopan, Sclater's monal, and Hume's pheasant; and the green peafowl is 'endangered'. Most of the others are also affected, especially on account of habitat loss and degradation in recent years, and also because they have restricted or fragmented ranges. Scientific status surveys for these species are urgently required (see Appendix 1, 3 and 4).

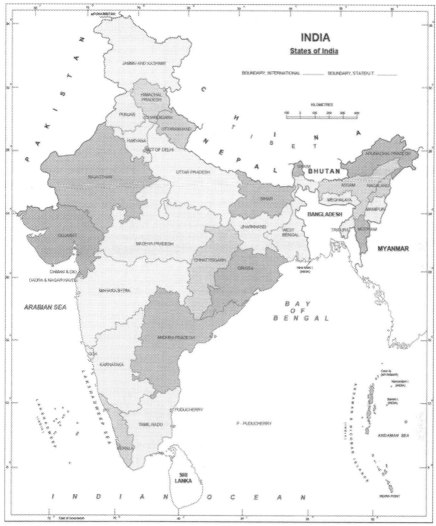

Fig 6.1 A map showing the states of India.

The Galliformes are important indicator species and their presence in an area is a good indicator of the health of the ecosystem. As in the case of most wildlife, the single most serious threat to the survival of Galliformes is the loss of forest and grassland habitat, and its degradation and fragmentation. The other major threats are human disturbance, killing or trapping for local consumption, and the use of pesticides and other chemicals in agriculture.

The initiative for the conservation of pheasants and other Galliformes in India is quite recent. In Europe and USA, the captive breeding and aviculture of introduced ring-necked pheasants *Phasianus colchicus* has been taken up in a big way and several pheasant farms have been established, where millions of pheasants are raised every year. The underlying motivation is sport and hunting, which is responsible for giving to these beautiful birds the tag of 'game birds'. The first serious attempt to orient pheasant rearing activity towards conservation was made when the Pheasant Trust was formed in the UK in the late 1950s. This paved the way for the establishment of the World Pheasant Association (WPA) in the mid-70s. Thereafter, the WPA helped establish regional chapters, and the one in India, called the WPA-India, was set up in 1979. Since then, in collaboration with WPA International, WPA-India has been striving to promote the cause of pheasants and other Galliformes in the country. WPA-India aims at creating awareness in the country about the ecological, economic, and aesthetic importance of Galliformes in general and pheasants in particular, and works to generate support for the conservation of these birds from government and non-government bodies and individuals. WPA-India also carries out research on individual species in their natural habitats.

One major achievement in the working of WPA-India has been networking and partnerships with other agencies and organisations. Collaborative relationships have been forged with the Wildlife Institute of India, Central Zoo Authority, National Zoological Park, Bombay Natural History Society, Centre for Environment Education, Aligarh Muslim University, and others. In April 2004, the 3rd International Galliformes Symposium was held in India, and the venue was the Wildlife Institute of India (WII) at Dehradun (in Uttarakhand). An important output of this event was the unanimous adoption of the South Asia Strategy for the Conservation of Galliformes in the region. Subsequently, a Memorandum of Understanding was signed between the WII and WPA-India - aimed at undertaking and facilitating research, training, and educational activities for promoting the conservation of Galliformes.

Many populations of Galliformes species and their habitats have been protected through the Protected Area (PA) network of India, but substantial populations of them still occur outside the PA network. For instance, the distribution range of pheasants such as Tibetan eared pheasant *Crossoptilon harmani*, Sclater's monal *Lophophorus sclateri*, green peafowl *Pavo muticus*, grey peacock pheasant *Polyplectron bicalcaratum*, Blyth's tragopan *Tragopan blythii*, and Temminck's tragopan *Tragopan temminckii* are not adequately covered under the PA network in the eastern Himalayas; and the same is the case for western tragopan *Tragopan melanocephalus* and cheer *Catreus wallichi* in the western Himalayas. To protect the habitat of these species, their habitats have to be declared as PAs, Conservation/Community Reserves or Biodiversity Heritage Sites. Lack of basic information such as the status and distribution, and researched information on the population ecology, habitat use, and behavior, has been a limitation in planning conservation of Galliformes in India.

During the last two decades, there have been surveys and studies on Galliformes in India that have resulted in a better understanding of some of the species status and distribution, and have helped in planning appropriate *in situ* and *ex situ* conservation efforts. The Wildlife Institute of India (WII) has been working in the field of Galliformes research and training since 1989. These include nation-wide surveys for the assessment of the status and distribution of the Indian peafowl and the red junglefowl, surveys for pheasants in the western and eastern Himalayas, the ecology and conservation of pheasants in the Himalayas and of the Nicobar megapode. WII has also conducted training of field staff in monitoring of Galliformes in the Himalayas, at the State and Division levels.

In much of rural India, it is possible to sight species such as the blue peafowl, black partridge, red jungle fowl, or kaleej pheasants, species that are associated with human edge habitats. The rural tourism industry and home-stays rely much on bird-sightings for attracting domestic and international tourists. Birds are an important part of the food-chain in agricultural ecosystems, and contribute to it by controlling insect pests, aiding seed dispersal, pollination, being a food source for hunting communities, and even scavenging and keeping the ecosystem free from pathogens. The resident birds as well as migratory waterfowl and raptors that visit the rural countryside are traditionally protected by many rural communities. Also, after the 74th Amendment to the Indian Constitution in 1992, which places the responsibility of conserving the urban environment with the local municipalities, efforts are being made by the local municipal governments to preserve wildlife habitats in and around urban areas through municipal notifications. Such habitats in the cities, like wooded areas in institutional campuses and cantonments, wetlands and riverfronts, provide a refuge to the wild birds and animals, as well as facilitate educational and recreational efforts for the city dwellers (Chauhan, 2013).

In this chapter, conservation activities being undertaken in the states of India are highlighted, with a detailed picture of the activities in the western Himalayan state of Himachal Pradesh, the leading state in pheasant conservation.

A Brief Introduction to Himachal Pradesh

Himachal Pradesh is one of the states in the western Himalayas in India. The state is surrounded by the states of Jammu and Kashmir in the north, Punjab in the west, Haryana and Uttarakhand in the south, and has a border with Tibet in the east. The state geographically consists of 3 mountain ranges of the Himalayas – the Himalayan foothills (Dry sub-tropical to temperate alpine forests), Greater Himalayas (Moist temperate alpine forests), and the Trans Himalayas (cold and arid rain-shadow area east of the Greater Himalayas, consisting of rugged treeless mountains and steppe-like grasslands).

Himachal Pradesh is divided into 12 districts. Its capital city is Shimla, also famous as the 'summer retreat' of the British colonial era, since it served as the Summer Capital of India, and the administration of the country shifted to this hill town during the oppressively hot Indian summers. The State has a rich cultural history, the predominant influences being the Rajputs, the Mughals, the Sikhs, the Tibetans, the Gurkhas, and the British. The primary occupation of the people is agriculture, horticulture, pastoralism, and tourism. The State has a substantial rural, tribal, and nomadic-pastoral population, and yet has a literacy rate of 84%, amongst the highest in India. The main industries in the State are cement manufacturing, hydel-power, tourism, IT, and food processing. The State is rich in natural resources. The major rivers of the State – Chenab, Ravi, Beas, and Sutlej are tributaries of the River Indus; and Tons, Giri, and Pabbar are tributaries of the River Yamuna.

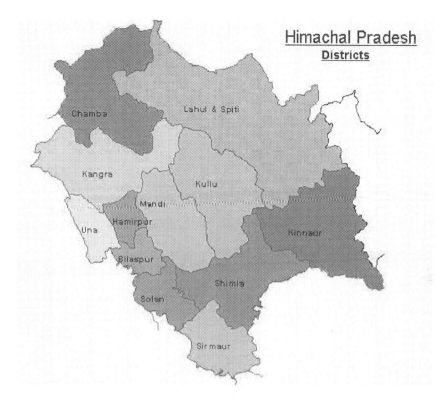

Fig. 6.2 A map showing the 12 districts of the State of Himachal Pradesh.

Traditional handicrafts (paintings, wood-carvings, woven shawls etc) from Himachal Pradesh are recognized all over the world, and some of these products, like the Kangra tea (a mixture of 20 herbal ingredients including green tea, from Kangra district), Chulli tel (bitter almond oil prized for use in cosmetics and pharmaceutical industries, from Kullu district), and Kinnauri shawls (from Kinnaur district), also have the Geographical Indication Tag as these are popular traditional products.

There are many places of interest for tourism in the State. Some of them are the Hindu temples, the Buddhist monasteries, and the British colonial era monuments. The Kalka to Shimla railway line, constructed by the British, is a UNESCO World Heritage Monument. The railway passes through some of the most picturesque scenery in the Himalayas, and is dotted by over a 100 tunnels through its 6 hour journey. One of the national parks in the State – the Great Himalayan National Park in the Greater Himalayas - has been nominated as a World Heritage Site by UNESCO due to its unique and rich biological diversity and geological beauty. The State also has 33 wildlife sanctuaries. Almost 26 % of the State's geographical area is covered in forest. It supports 7 of the 17 pheasant species found in India. The State Animal, Bird, and Flower are the snow leopard, the western tragopan, and the pink rhododendron, respectively.

Protected Areas in Himachal Pradesh

Fig 6.3 A male koklass pheasant walking along the road in the Shimla Water Catchment Sanctuary, Himachal Pradesh, India. Photo by Anita Chauhan, 2010.

The forests in Himachal Pradesh are an important wildlife habitat in the western Himalayas. A network of 33 Wildlife Sanctuaries and 2 National Parks constitute Himachal Pradesh's Protected Area Network (PAN) that contains its rich biodiversity. It includes several globally threatened species like snow leopard, Himalayan ibex, musk deer, western tragopan *Tragopan melanocephalus,* and cheer pheasants *Catreus wallichii.* This network (PAN) comprises 13.6% of the state's geographical area. The PAN shall further increase to 15.11% after the ongoing 'rationalization' process of existing National Parks and Sanctuaries is completed. In the rationalization process, the boundaries of protected areas are re-drawn, so as to exclude villages and those forests that are heavily laden with people's traditional rights of harvesting timber, fuel wood, grasses, etc. This exercise will address the demands of the people concerning their traditional rights and also help in implementation of development schemes. In February 2003, HP State Government declared the panchayat pradhans (elected village heads) as forest officers; and has listed 38 items of Minor Forest Produce that can be extracted from the protected forests after paying a fee, or after obtaining permits from them. This will put the responsibility of sustainably harvesting the forest upon the villagers' shoulders. The permits for extraction of MFP are also given by the Divisional Forest Officers (DFO). Moreover, new ecologically rich, rights-free forest areas contiguous to the existing protected areas shall be added to the network. After the rationalization of forest areas is completed, the PAN will comprise 5 National Parks, 26 Wildlife Sanctuaries and 3 Community Reserves. This exercise will provide a major boost to the conservation of wildlife.

The Galliformes (particularly pheasants and francolins) have been hunted by villagers in the Himalayas to varying extent to supplement their diet, and for the sale of meat and feathers, since time immemorial. Historical literature reveals that hunting of kalij pheasants, chukor *Alectoris chukar*, black francolin, and mammals, such as barking deer and goral, was common. During the British rule in India, heavy hunting of birds and mammals, and large-scale deforestation, caused a drastic fall in animal populations, and many a species became quite rare. In recent times, hunting pressure, accompanied by habitat fragmentation, rapidly increasing human population, fires, droughts, climate change, etc., has been responsible for further decline in wildlife populations. A complete ban on hunting has been imposed since 1990 (Dhiman, 2011).

Himachal Pradesh is home to 7 species of pheasants, namely – the western tragopan, Himalayan monal, cheer, koklass, white-crested kalij, red junglefowl, and blue peafowl, found in GHNP and 27 Sanctuaries in the PA network, and also in forests outside the network. Though data for the total abundance of pheasants in the state is not available yet, methodical surveys done in some habitats harboring pheasants have helped in assessing their status. These surveys have been conducted in the upper Beas valley, parts of Ravi and Sutlej valleys, and some parts of the Yamuna catchment. As mentioned in chapter 3, 'long term ecological monitoring' plan for GHNP is being prepared by the H.P. Forest Department to help with the regular monitoring and management of biodiversity.

Table 6.2 27 of the 33 Sanctuaries in the state have various species of pheasants (Gulati et al., 2004).

Pheasant Species	Name of the Sanctuary
Western tragopan	Daranghati, Sainj, Tirthan
Koklass	Bandli, Chail, Churdhar, Daranghati, Dhauladhar, Gagul Siyabehi, Kalatop Khajjiar, Kanawar, Khokhan, Kias, Kugti, Manali, Nargu, Renuka, Rupi Bhabha, Sainj, Sechu Tuan Nala, Shikari Devi, Shimla Water Catchment, Tirthan, Tunda
Cheer	Chail, Daranghati, Dhauladhar, Kugti, Lippa Asrang, Majathal, Rakchham Khitkul, Sainj, Tirthan
Himalayan monal	Churdhar, Daranghati, Dhauladhar, Gagul Siyabehi, Kanawar, Khokhan, Kias, Kugti, Lippa Asrang, Manali, Rakchham Khitkul, Rupi Bhabha, Sainj, Sechu Tuan Nala, Shikari Devi, Tirthan, Tunda
Kalij	Bandli, Churdhar, Daranghati, Dhauladhar, Gagul Siyabehi, Kalatop Khajjiar, Kanawar, Khokhan, Kias, Lippa Asrang, Majathal, Manali, Nargu, Rakchham Khitkul, Rupi Bhabha, Sainj, Sechu Tuan Nala, Shikari Devi, Shimla Water Catchment, Tirthan, Tunda
Red junglefowl	Chail, Gobind Sagar, Majathal, Pong Dam, Renuka, Shimla Water Catchment, Simbalbara
Blue peafowl	Chail, Gobind Sagar, Pong Dam, Renuka, Simbalbara

Hunting, habitat fragmentation, and encroachment by people and livestock remains a problem in HP; enforcement in remote areas is difficult due to lack of infrastructure. Conservation action in terms of education programs, alternate livelihood - cultivation of medicinal herbs, and employment in tourism industry, etc - has been initiated for people living close to the Protected Areas. This has proved effective in reducing the stress on the ecosystem in GHNP, and the western tragopan numbers have shown an improvement in the census conducted in year 2010.

Ex-situ Conservation - Zoos and Conservation Breeding Projects

As discussed in Chapter 4, individuals from *ex-situ* populations can be periodically released into the wild to maintain numbers and genetic variability in natural populations. Captive breeding also helps in the acquisition of scientific knowledge about the behavior and other aspects of the species, as well as serving as an educational tool.

Himachal Pradesh has recognized pheasants as flagships of their forests and has initiated projects to conserve them. Three *ex-situ* breeding projects approved by the Central Zoo Authority are in different stages of progress in Sarahan (for western tragopan), Manali (for Himalayan monal) and Chail (for cheer pheasant). Though chicks have now been hatched in *ex-situ* conditions in the pheasantries, the actual re-introduction is a long way away. Simultaneous projects to identify sites for re-introductions are being planned, so that managed sites are available for release. The Himachal Pradesh Forest Department officials and scientists have also been working on forming 'Husbandry Guidelines for Cheer Pheasants' based on the IUCN guidelines, for the breeding project of this species.

Some of the zoos in the state also house pheasants, the most popular among them is the Kufri Nature Park in Kufri, near Shimla. Zoos and pheasantries serve as a means to familiarize the general public, school and college students, and tourists about the pheasant species.

Fig 6.4 The pheasantry at the Himalayan Nature Park in Kufri, Himachal Pradesh houses 8 species of pheasants. Photo by Satpal Dhiman.

It will be helpful to have a historical perspective on how conservation projects have started and evolved in Himachal Pradesh, from the first attempts in the 1980s. Himachal Pradesh (which literally means 'abode of snow'), Punjab, and Haryana, were earlier a combined state (with Chandigarh as the capital city), but Himachal was formed as a separate state in the year 1971. 'Also during this time in India, the Wild Life (Protection) Act, 1972 (WPA, 1972) was enacted. The WPA, 1972 provided a greater impetus to the setting up of the system of National Parks and Protected Areas in India. The new Government of Himachal Pradesh was interested in establishing one or more national parks in the State. The first Chief Minister of the State, Mr. Y.S. Parmar, recognized the need to protect the State's wealth of wildlife species. Mr. Dilaram Shabab, the legislator from the Tirthan valley, had already been discussing possibilities for setting up of wildlife Protected Areas with the Chief Forester of the State. Mr. Shabab wrote to Mr. Parmar in 1971, proposing the Sainj-Tirthan valley region, up to its boundary with the Spiti region, as a new national park, both to conserve the wildlife, and as a development strategy for nearby villages. So, the newly set up Himachal Forest Department, was now considering the establishment of Protected Areas. Momentum was rising for establishment of the Great Himalayan National Park' (Tucker, 1997).

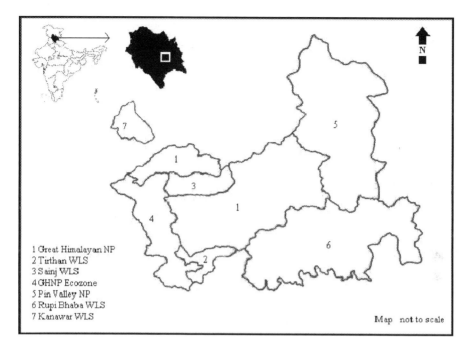

Fig. 6.5 Map showing the location of Great Himalayan National Park, and the surrounding Protected Areas in Himachal Pradesh, India.

In 1976, 8,396 hectares were declared as the Tirthan Wildlife Sanctuary. In 1978-80 a team of Indian and international wildlife biologists (A.J. Gaston, M.L. Hunter, and P.J. Garson) conducted the first systematic survey (the Himachal Wildlife Project) of 3 areas in the upper Beas region as possible sites for a national park. In 1981, their report recommended the Sainj-Tirthan valley area as having less human and livestock impact than either the Solang valley at the upper end of the Beas basin or the upper Parbati valley. In March 1984, the Himachal Government declared its intention to create GHNP in that area; and after another survey by Garson and Gaston in 1991 (that recommended regulation of cattle grazing, and medicinal plant collection in the area), the State Government created the Great Himalayan National Park and an Ecodevelopment Zone on its western side, as well as created the Sainj Sanctuary adjoining the Park (Tucker, 1997). The Pin Valley National Park, located in the cold desert in the rain-shadow area in Spiti valley, was also established in 1987.

Since its formation in 1975, the World Pheasant Association has focused on ecological conservation of pheasants, and on aviculture as a means of promoting the pheasants' survival. Major conservation initiatives undertaken during that period include detailed survey work on two species then considered to be among the most threatened - the cheer pheasant *Catreus wallichi* and the western tragopan *Tragopan melanocephalus*. Both occur in the western Himalayas, the cheer on mid-altitude grassy slopes with scattered rocky outcrops interspersed with pine trees and gullies, and the tragopan in moist temperate forests. The cheer pheasant was viewed with concern because it seemed to be widely scattered in tiny populations in habitats severely affected by human activities. Extensive surveys through much of the 1980s revealed that the species seems to thrive in early succession vegetation that is regularly grazed, cut for fodder, or burned (McGowan, 1994).

Subsequent conservation efforts for the western tragopan in Pakistan, have taken the form of the ambitious Himalayan Jungle Project, led by Birdlife International, with the objective of conserving an extensive tract of pristine forest in the Palas Valley, by encouraging the local people to use the forest sustainably, which will ensure long-term benefits, while allowing the western tragopan to survive in this unique habitat. Ironically, the prohibition of grass-cutting and grazing in Margalla Hills National Park in Pakistan, over a decade or so, led to the growth of scrub cover, which makes it unsuitable for the cheer, though it supports a good population of kalij pheasants (McGowan, 1994). This presents an additional management aspect to the scientists who are today exploring alternative sites for the reintroduction of cheer pheasants in Himachal Pradesh. 'This species is also the only one in the pheasant family which has been the subject of a concerted reintroduction attempt, started in the Margalla Hills National Park in Pakistan. The natural cheer population became locally extinct in 1976 in this National Park. Eggs from some of the large number of captive cheer in Europe were flown to Pakistan, where they were incubated and reared, and subsequently released. Radio-tracking has shown that released birds can survive in the wild for as much as 18 months. The mass rearing of chicks largely or entirely in the absence of adult birds produced poults that roosted on the ground at night and were generally very prone to predation. Despite about 15 years of effort, there was no evidence of a self-sustaining wild population being established. Later, when the first surveys of cheer populations were conducted in the Indian state of Himachal Pradesh, it was also found that most cheer pheasants occurred at altitudes above that of the release site at Margalla. The cheer project has, nonetheless, resulted in the formation of a conservation society in the Margalla Hills National Park, which promotes public awareness about the value of wildlife resources' (McGowan, 1994). Studies conducted on other Galliform species have also shown that the presence of the mother is essential for the normal behavioral development of chicks, and helps them avoid predators. The lessons learned from such attempts are important for fine-tuning the re-introduction technique of pheasants.

Cheer Pheasant Conservation Breeding Programme

Fig 6.6 Photo of a pair of cheer pheasants in Uttarakhand, India by Kousik Nandy, 2012.

A restricted range species, the cheer is distributed through the southern foothills in the western Himalayas from north Pakistan, through Kashmir, into Himachal Pradesh, and Uttrakhand (India), and east to central Nepal. Its distribution is patchy as it occurs in small isolated pockets in sub-tropical or lower temperate grasslands/scrublands. Majathal and Chail Wildlife Sanctuaries in Solan district, and GHNP in Kullu district are the best recorded sites in Himachal. Past records also indicate their presence in areas of Chamba, Kinnaur, and Shimla districts. In the spring of 2008, two important areas have been found in a field survey by Dr. K. Ramesh from WII, at Pilang in Uttarkashi District, and Namik in Pithoragarh District of Uttarakhand State in India. The Pilang valley is sparsely populated and rich in Galliformes, including a good cheer pheasant *Catreus wallichii* population. The Namik valley is flanked by other areas with good forest cover, and together they may host one of the largest populations of satyr tragopan *Tragopan satyra* in India. Proclamation of 'Conservation Reserve' or 'Community Reserve' status for the two areas has been proposed to the State Forest Department. These relatively new categories of protected areas involve the sustainable integration of conservation and human livelihood considerations. Cheer pheasant is listed as 'Vulnerable' in the IUCN Red List.

Fig. 6.7 Map showing the distribution of cheer pheasants in India, 2007.

Keeping in view its declining natural populations, a captive breeding facility for cheer pheasant was created at Blossom, Chail, Himachal Pradesh in 1987. However, there was a lack of knowledge about the breeding and ecological behavior of the species. Field studies were undertaken by the H. P. Forest Department staff to understand the needs of this species. Breeding pairs in the cheer sites located in the adjacent Chail Wildlife Sanctuary were studied. The ecological and behavioral needs of the captive stock were then addressed, and the pheasantry was successful in breeding this species in year 2002, though problems due to inappropriate and small enclosure sizes, inbreeding, and high juvenile mortality were major bottlenecks (Dhiman, 2011).

In the new conservation breeding project, the enclosures have been scientifically designed and constructed to remedy the problems present in the old enclosures. The old pheasantries (at Blossom and Khariun) are rectangular in shape, and have 6 to 8 small rectangular enclosures with an area of about 36 m² available to each breeding pair (Dhiman, 2011).

Cheer pheasants eat a variety of food in the wild, including seeds, berries, grasses, roots and tubers, and insects. In captivity they are fed grain mixture, wheat and barley grass trays, vegetables, fish meal, soya meal, and limestone. Any vegetation planted in the small enclosures to supplement their diet was dugout and demolished by the cheers, as they are excellent diggers, and eat roots and tubers. Pheasant food pellets used in European pheasantries can provide a balanced diet for better breeding success, but pellets do not encourage natural feeding behavior (Dhiman, 2011).

Fig 6.8 A cheer pheasant family – a male, a female, and 2 chicks - in the older enclosures at Khariun, Chail. Photo by Satpal Dhiman.

In the old enclosures, predators such as yellow throated martens, weasels, rats, and snakes, posed a great threat to the pheasants. The wire mesh size used in these old enclosures was not suitable to keep snakes and rats out. Martens sometimes invaded the enclosures to prey on adults; snakes and rats ate eggs and the chicks. The problem was temporarily addressed by putting an extra chicken wire mesh. Also, juvenile mortality data collected between year 2005-2009 indicates that significant mortalities have been caused by injuries/trauma and predation. The newly hatched chicks in these facilities died because of drowning in mud puddles created in the lower end of the sloping enclosures (Dhiman, 2011).

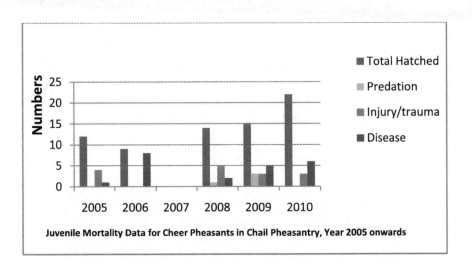

Fig. 6.9 A graph showing the juvenile mortality data for cheer pheasants in Chail pheasantries, year 2005-2010, by Satpal Dhiman.

The 7 new hexagonal enclosures constructed at Khariun have an area of about 225 m² available to each breeding pair. This is perhaps the biggest ever enclosure designed for the conservation breeding of any pheasant species in the world. Large enclosure size is considered vital for a cheer breeding pair so that they can rear their large brood in near-natural environment. In the new enclosures, galvanized wire mesh and chicken wire mesh have been fitted in a metal frame. The wall foundations are rodent-proof. Adjacent aviary walls are higher and tapering on the top to prevent fighting between males during the breeding season. The enclosures have been designed to provide nesting spaces and have been planted with natural feed. Water-proof roofed night shelters are provided for roosting and for refuge during inclement weather (Dhiman, 2011).

Fig 6.10 A. A view of the hexagonal pheasant enclosures at Khariun, Chail, Himachal Pradesh, B. Inside the new hexagonal enclosures at Khariun, Chail. Photos by Satpal Dhiman.

Important steps taken for the management of the captive cheer include identification and marking of founders, and preparation of individual bird sheets. This breeding project endeavors to address the conservation needs of the species, and complement the *in-situ* conservation efforts (Dhiman, 2011).

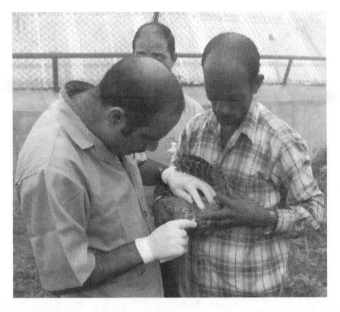

Fig 6.11 A veterinary doctor examining a cheer pheasant in the Khariun pheasantry, Chail, Himachal Pradesh. Photo by Satpal Dhiman.

The Western Tragopan Conservation Breeding Programme

Western tragopan is the state bird of Himachal Pradesh. ['Tragus' means 'goat' in Greek, and 'Pan' is the name of a god of shepherds and flocks that is worshipped in ancient Greece. The bird is so called because its call sounds like that of a goat, and it has two fleshy 'horns' about 3 inches in length on its head that are visible only during its courtship display dance]. The bird is revered among the local people, and finds a place in folk songs. According to one legend, after God had created all the birds, He took a sliver of colour from every bird there was, and created the magnificent tragopan – locally called *Juju Rana*, or 'King of Birds'. The bird is so elusive that it has only recently (in June, 2010) been photographed in the wild for the first time (in Great Himalayan National Park), by India's ace photographer Dhritiman Mukherjee. This species inhabits a relatively small range in the western Himalaya, of mixed coniferous-broadleaf forests with ringal bamboo in moist temperate and sub-alpine zones. Till 1980s, such areas were decreasing at an alarming rate due to deforestation. The bird is listed as 'Vulnerable' in the IUCN Red List.

Fig 6.12 A western tragopan male at the Himalayan Nature Park at Kufri, Himachal Pradesh, India. Photo by Anita Chauhan, 2009.

6.13 A map showing the distribution of western tragopan in India, 2007.

Its global distribution is limited to parts of Pakistan, Jammu and Kashmir, Himachal Pradesh, and Uttarakhand. In Pakistan, it survives only in a few isolated pockets. Presently there is no record of a definite sighting in Uttarakhand. In Jammu and Kashmir, it has been recorded in Neelam and Jhelum river valleys, and in undisturbed forested areas of Kishtawar and Doda. In latest surveys in Himachal Pradesh, this species is recorded at Great Himalayan National Park (Kullu district), Rupi-Bhaba Sanctuary (Kullu district), and Daranghati Sanctuary (Shimla district). In the past, it has also been recorded in Kangra, Kinnaur and Chamba districts.

Records made during a survey along the Pir Panjal range in May 2011 have confirmed its presence at two new sites in Jammu and Kashmir. Sightings and calls of the pheasant were validated at the Kalamund-Tatakuti and Khara Rakh areas of the range, by Riyaz Ahmad, the team leader and Assistant Manager, Species Division of Wildlife Trust of India. In addition to western tragopan, the team also sighted the cheer pheasant (*Catreus wallichii*). Unlike its usual haunts, the moist north facing coniferous slopes, the present sites are located on the south face of Pir Panjal along Poonch. The next step would be to gauge what numbers exist at these sites to effect conservation. The team has recommended Kalamund-Tatakuti for notification as a Protected Area. A victim of rampant poaching for its meat and plumage and habitat degradation and fragmentation, the western tragopan has previously been reported only from Kazinag range and Kishtawar National Park in the state. A few scattered records occur from Sud Mahadeo area of Jammu province (Wildlife Trust of India, 2011).

Unlike the cheer, the western tragopan does not readily reproduce in captivity. The pheasantry at Sarahan Bushahr in Shimla district is famous for being the only pheasantry in the world where the western tragopan has bred in captivity. Sarahan Pheasantry was established in the year 1987 with the objective of housing rescued pheasants and other animals driven down from the forests to human habitations during winters in search of food.

The infrastructure at the pheasantry includes guard huts, chowkidar huts, store rooms (for feed/medicines), and 1 quarantine unit. The entire area is fenced to safeguard the pheasants from predators. The pheasantry has 28 pheasant enclosures which are constructed using wooden frames and wire meshes. The average size of the pens is about 30 m². The small size of the enclosures poses a problem for free movement and display during breeding season and for raising the clutch of young ones.

There are 19 western tragopans (10 females and 9 males) at this pheasantry along with 9 male monals, one female koklass pheasant, and a few kalij pheasants. Koklass, monals, and kalij pheasants are housed in a circular pheasantry, and the 8 breeding pairs of western tragopans are housed in rectangular enclosures constructed in series. In the past, the pheasants have reproduced successfully at this pheasantry and have been sent to various other zoos in Himachal and other states.

Table 6.3 Hatching data of western tragopan at Sarahan pheasantry.

Year	No. of Chicks	Mortality
1993	1	-
2005	2	1
2007	10	1
2008	6	2
2009	2	-
2010	1	1
2012	7	-

In the year 2005, a plan for the construction of a new Conservation Breeding Center for the western tragopan at Sarahan was put forward by a core group of scientists, wildlife administrators, members of the Central Zoo Authority, the WPA, and other NGOs. CZA approved a monetary grant for construction of 7 enclosures at a location near the present pheasantry. The size of the new enclosures per breeding pair will be 280 m². The larger enclosures will provide a near-natural habitat for the breeding pair, and will allow adequate display area to the male during the breeding season, as well as accommodate the hatched brood of young ones. The infrastructure at the new captive breeding center will include 7 enclosures with perches, night shelters and vegetation, a clinic, a store, a kitchen, etc.

Fig 6.14 A. A female western tragopan, and B. a western tragopan chick at the Sarahan pheasantry, Shimla district, Himachal Pradesh, India. Photos by Satpal Dhiman.

A Visit to the Sarahan Pheasantry

Sarahan Bushahr is a small township in the Shimla District of Himachal Pradesh, India. It is famous, as much for the grandiose Bhima Kali Temple, as it is for the Sarahan Pheasantry. It is situated at an altitude of 2310 m and is a 6hours drive from Shimla via the township of Rampur Bushahr. The National Highway 22 (Shimla-Kinnaur) passing through Rampur, branches off after Jeori, from where a road climbs up to Sarahan. The distance from Rampur is 50 kms and the route passes through small towns - Jhakri, Jeori, Kotla, Gharat, and Bonda.

Sarahan town is located on a giant ridge about half-way down perennially snow clad mountains. The River Satluj flows along the base of the mountain. The entrance to the pheasantry is located near a sports stadium. After a 15 minute walk from the entrance gate, up a mountain trail carpeted with pine needles and cones, the visitors reach the ridge on which the pheasantry is situated. This world famous pheasantry is the only place the western tragopan has reproduced in captivity.

A signboard informs visitors about the Dos and Don'ts while they are in the pheasantry. The first enclosure is the circular pheasantry which houses magnificient monal, koklass, and kalij pheasants. The male monals glisten as the noon sunlight bounces off their silken plumage. The kalijs can be seen whirring their wings and dust-bathing. Scores of kalij hatched at this pheasantry have been sent to other zoos in the country in the past. There are concrete benches and a gazebo where visitors can rest and enjoy the beautiful view of the Bhimakali Temple and the snowcapped mountains. A few meters away, there are several rectangular enclosures which bear the names of the individual western tragopans housed in them. The bright crimson color of the male's plumage helps one spot the bird in the shadows of the enclosures.

The birds can be seen perched on a branch or pecking away at their meal trays. The loud call of the ubiquitous Himalayan great barbet provides constant background while one is busy spotting delightful turquoise verditer flycatchers and various species of tits visiting the nearby trees. One can also spot some rock agamas sunning themselves on the rocks just off the edge from the gazebo.

Fig 6.15 A. The circular pheasantry pen and gazebo, and B. rectangular enclosures at Sarahan Pheasantry, Himachal Pradesh. Photos by Anita Chauhan, April 2011.

Fig. 6.16 A. Kashmir rock agama sunning, and B. Verditer flycatcher, in Sarahan pheasantry, Himachal Pradesh. Photos by Anita Chauhan, 2011.

New scientifically designed enclosures, for a conservation breeding center for western tragopan, are being built at a location near the present pheasantry. The western tragopan breeding pairs will be shifted to the 7 new enclosures.

The pheasantry staff, which includes the caretaker Mr. Alam Singh, Forest Guards, and workers, is always at hand in case the visitors need any information. After the visit to the pheasantry, visitors can walk down to Sarahan town to see the magnificient Bhīma Kali temple, and buy souvenirs at the market. Sarahan town also has hotels and eateries for tourists.

Daranghati Wildlife Sanctuary, in the Rampur Bushahr division of Shimla district (Himachal Pradesh, India) is located a short distance away from Sarahan. A former hunting reserve of the Rampur Bushahr royal family, it was notified as a sanctuary in March 1962. The sanctuary is in two non-contiguous sections namely, Daranghati Wildlife Sanctuary I and II, on either side of the Dhaula Dhar mountains, a range that forms a part of the Greater Himalayas. The total notified area is 168 km2; and the altitude varies between 2,100 m and 5,400 m (Singh and Tu, 2008). Four pheasant species – the western tragopan, koklass, Himalayan monal, and kalij – are found in good numbers in this Sanctuary. The DFO's office in Sarahan can be contacted for permission to visit the Sanctuary.

Fig. 6.17 Bhima Kali Temple, Sarahan, H.P. Photo by Anita Chauhan, 2011.

The pheasantry is closed during the pheasant breeding season from 15th April to 31st July.

Himalayan Monal Conservation Breeding Programme

The Himalayan monal *Lophophorus impejanus* is found throughout western and central Himalayas, from Afghanistan in the Safed Koh to Bhutan, being replaced in the east by Sclater's monal *Lophophorus sclateri*. The species is not listed in the 'threatened categories' of the Red List. Some of the areas where monals occur are GHNP, Kanawar, Rupi-Bhaba, Daranghati, and Talra Wildlife Sanctuaries. Systematic surveys in GHNP reported monal encounter rates ranging between 1.5 to 3.9 birds/km walk. Surveys in the PAs of Kanwwar, Rupi-Bhaba, Daranghati, and Talra suggest 3-7 birds/km².

Monal is a conspicuous pheasant that is very well known to the local villagers, and its calls are often recorded by trekkers. The monal is found throughout the middle Himalayan range between 2,000 and 3,500 m, depending upon the season. Locally, it is known is *Bnal* (male), *Bodh* (female), or *Kardi* (female), in Shimla, Kullu, and Pandrabis areas; and *Dang* in Kinnaur district. The local people consider it chivalrous to adorn their caps with a male monal's crest, which is made of wire-like spatula tipped feathers and skin. In Kullu district, population density of calling male monal has been recorded between 1-10/km². The best sightings of monal have been made in Shimla hills and the upper Beas Valley between 2,600-3,150 m. However, the front ranges of Himalayas nestling the Shimla Water Catchment area, which during British time, had monal populations, are now devoid of this pheasant. The nearest monal reported in this area is from Hatu peak above Narkanda in Shimla district.

A conservation breeding center for the Himalayan monal is being planned at Manali in Himachal Pradesh.

Conservation of Pheasants in the other States of India

As mentioned earlier, 16 of the 17 Indian pheasants are found in the Himalayas. The blue peafowl and the red junglefowl are found in almost all the States (Himalayan and Peninsular) of India. The status, distribution and conservation of these 2 species will be described first, followed by a brief summary of pheasant conservation in the States of India (Sathyakumar and Sivakumar, 2007).

Blue Peafowl

The blue peafowl *Pavo cristatus* has been an integral part of Indians and their culture for centuries. From religion and mythology, the peafowl occupies an important place in the lives of the people. In addition to this, the blue peafowl is recognized for its ecological and aesthetical values, and was hence declared as the 'National Bird' of India in the year 1963.

Although the Indian- or blue peafowl is widely distributed, and locally abundant, or fairly common in some areas, the present population status of this species is only speculative. Most of the Protected Areas (PAs) in India do not regularly monitor the populations of the Indian peafowl within their areas despite this species being the 'National Bird', and even if it is an important prey species for carnivores such as tiger and leopard.

There is an urgent need to obtain basic information on blue peafowl presence/absence, encounter rates and population estimates from the PAs, outside PAs, including revenue and private lands, for the better management of the National Bird. It appears that substantial portions of blue peafowl distribution range and populations are outside the PA network or Forest land areas. According to a recent questionnaire survey by WII, the blue peafowl presence is confirmed in 345 districts of India and it is likely to be present in 174 other districts that fall within the blue peafowl distribution range. The questionnaire survey also found that the blue peafowl forms one of the important preys for large and small carnivores, both within and outside PA network or forest land areas. Blue peafowl is predated by wild carnivores such as tiger, leopard, leopard cat, jungle cat, caracal, wild dog, wolf, jackal, bear, fox, and mongoose.

As a follow up of the survey, WII proposes to coordinate a national level blue peafowl Monitoring Programme, networking with 5 regional institutions in the country, NGOs, NGIs, and the State Forest/Wildlife Departments. WII also proposes to use and analyse the field based data generated by Phase-I of the "All India Tiger, Co-predators, Prey and Habitat Monitoring Programme", in which information on the presence/ absence and relative abundance of the blue peafowl at the Beat Level for all the 17 Tiger Range States has been collected. This could serve as the baseline for regular monitoring of blue peafowl in the tiger landscapes of the country.

Since the early 1990s, there have been reports of increasing illegal trade in peafowl feathers, large-scale mortalities due to increased use of insecticides/pesticides in agricultural lands, poaching, and retaliatory killings by people due to alleged crop depredation by peafowl. Several peafowl stronghold areas in the country are now faced with its declining population trend. States that are the highest users of agricultural pesticides, such as Andhra Pradesh, Punjab, Karnataka, Tamilnadu, Maharashtra, Haryana, Gujarat, and Uttar Pradesh are reporting a high number of peafowl mortality caused by pesticides.

There are an alarming number of news reports of Indian peafowl mortalities in agricultural landscapes. The Indian peafowl is a granivorous bird and feeds on grains and seeds. This may result into conflicts and problems. The Indian peafowl is often labeled as 'agricultural pest' and if their impact on agricultural crop is adverse then their population may have to face retaliatory killings by local villagers (Choudhury et al., 2007). This has rendered it necessary to evolve a safe and practical method of crop protection from birds. At the same time one should not ignore the fact that the peafowl is also beneficial to agriculture because they destroy number of insect pests. Crops affected by peafowl include – paddy, ragi, sorghum, maize, groundnut, onion, and fruits. The farmers who raise paddy, maize, and groundnut in Tamil Nadu, admit that the birds cause damage to their crops by eating the grains and by the train feathers while moving about in the fields. But the farmers have not taken it as seriously, and often appoint boys to chase these birds away. The farmers are ready to give a small portion of their produce to these birds. Hindu mythology has given an exalted status to the Indian peafowl by associating it with Lord Muruga, which has helped their protection by local people. However, the poor farmers may not be in a position to accept crop damage by peafowl if it increases to higher levels.

As described in Chapter 4, the All India Network Project on Agricultural Ornithology (AINPAO) has developed certain bird scaring devices and recommended the same to the farmers for protecting their crops from bird damage. Reflective ribbon is used to scare the birds from crops such as sunflower, jowar, bajra, maize, guava and grapes, in an eco-friendly way. During the day, the light reflecting ribbons and their humming noise produced in the wind scares the bird from the field. The use of reflective ribbon in crop fields has helped in preventing crop damage by peafowl (Sathyanarayana, 2007).

Red Junglefowl

The red junglefowl *Gallus gallus* (RJF) is one of the most important species to mankind due to its economic and cultural significance. The present day multi-billion dollar poultry industry is based on the wild RJF, and may have to depend on it in the future as well. India is the 5th largest poultry egg producer in the world. Conservation of genetically pure wild forms has great potential to make a significant contribution to the study of some economically important genetic traits of the domestic form.

The RJF has widespread distribution and its 5 subspecies are spread across from the Indian sub-continent eastwards across Myanmar, South China, Indonesia, and Java. In India, 2 sub-species of RJF occur, the *Gallus*

gallus murghii and the *Gallus gallus spadiceus*. While the former is distributed in the north and central part of India, extending eastwards to Orissa and West Bengal, the later is confined to the North-eastern parts of India. Recently, fears had been expressed that the wild RJF populations may be genetically contaminated and there may not be any pure RJF populations in the wild (Fernandes et al., 2007). Contrarily, in a recent study by scientists, blood samples from 70 individuals of RJF (about half of them wild caught and the other half from zoo specimens) were subjected to molecular genetic studies, and the results showed presence of traits of the true red junglefowl.

6.18 Painting of red junglefowl *Gallus gallus* pair from *Indian Sporting Birds* by Frank Finn, 1915.

The conservation status of the red junglefowl is tied with the level of protection of the other pheasant species in western and eastern Himalayas. Although the arid and semi-arid zones in India have a better Protected Area network, there is immense pressure from human population growth and economic development, and these habitats face a huge challenge to balance human needs with the protection and management of natural habitats.

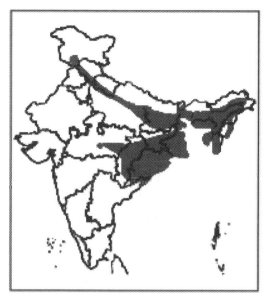

Fig. 6.19 A map showing the distribution of red junglefowl in India, 2007.

The main causes of threats are expansion and intensification of agriculture, infrastructure development and excessive livestock grazing. Clearance, conversion and degradation of habitats are by far the most important causes of threats to galliform species in these regions. Poaching is the 2nd most common category of threat. These species are poached for food and sport, and captured for the wild bird trade. Habitat destruction has led to the fragmentation of large continuous populations into small, isolated and scattered patches, rendering these species increasingly vulnerable to local extinctions. Large-scale and indiscreet use of pesticides have decimated large numbers of some species; blue peafowl being the worst affected in many locations in Rajasthan, Gujarat and Haryana.

Jammu and Kashmir

The state of Jammu and Kashmir is the northern-most state of India and its location and the range of altitudes provides for 3 distinct biomes – the lowland dry thorny scrub of Jammu, the mid level temperate evergreen forests of Kashmir, and the cold arid climes of Ladakh. The 3 regions also exhibit distinct faunal assemblages.

The state has 14 species of Galliformes, which includes 7 pheasants - Indian peafowl, red junglefowl, and the temperate species koklass, western tragopan, cheer, kaleej, and Himalayan monal. Only 2 of the 7 species, the western tragopan and cheer are threatened with extinction. All 7 species are included in the Wildlife (Protection) Act of Jammu and Kashmir state.

Status and distribution of galliformes in Jammu and Kashmir

1. Himalayan monal *Lophophorus impejanus*: It is a commonly occurring pheasant found in almost all Protected Areas (PA) within its distributional range in the state. It can be sighted in good numbers in Limber Wildlife Sanctuary (WS), Dachigam NP, Overa-Aru WS, and Kishtawar NP besides other forest areas of the state. No population estimates are available. The main threat in the past was hunting, but much of that appears to have been controlled now, although they are still trapped for food.

2. Western tragopan *Trogopan melanocephalus*: This threatened species is found at a handful of locations within Jammu and Kashmir. The largest population is perhaps in the Limber-Lachipora area of the Kazinag range. Western tragopan is also found in Kishtwar NP, and is reportedly present in Sud Mahadev area of Jammu province. Abundance estimates are available from Limber area only which suggests a population of over 45 males. They appear to be fairly secure in Limber area although nothing is known about their status from Kishtawar NP. Records made during a survey along the Pir Panjal range in May 2011 have confirmed its presence at two new sites in Jammu and Kashmir. Sightings and calls of the pheasant were validated at the Kalamund-Tatakuti and Khara Rakh areas of the range.

3. Cheer pheasant *Caterus wallichii*: The cheer has a limited distribution in the state. In Jammu area, it is confirmed from Trikuta area while in Kashmir it is found in the Limber WS and Lachipora WS. Southern slopes of the Pir Panjal range need to be surveyed to locate more sites. The species appears to adapt to disturbed habitats and therefore may not be as threatened as thought to be.

4. Koklass pheasant *Pucrasia macrolopha*: This is a relatively common pheasant (as in the other parts of the western Himalaya), and is found in most forested areas of the state. Most significantly, it is found in Overa-Aru WS, Limber WS, Lachipora WS, Gulmarg WS, Dachigam NP, and Kishtwar NP. Old records of the presence of this species from Boniyar area of the Kashmir valley are available. Population estimates for this pheasant are lacking.

5. Red junglefowl *Gallus gallus*: This species is mainly found in the Jammu region of the state and is found in the Ramnagar WS, Jasrota WS, and Nandini WS. They are also found in some Reserve Forests of Kathua.

6. White-crested kalij *Lophura leucomelanos*: This is also largely confined to the Jammu region of the state and is not found in any significant numbers in the valley, although some reports indicate that they may be present in the lower parts of the Jhelum valley. However this needs to be ascertained. The distribution largely follows that of the red junglefowl within the state.

7. Indian peafowl *Pavo cristatus*: This species occupies the areas adjacent to the Indian plains and is found in the Ramnagar WS, Nandini WS, and Nagrota WS in the state, besides near agricultural farmlands in Sambha and Kathua areas of Jammu. It is not found in the Kashmir Valley and Ladakh region.

The state is rich in Galliformes and a concerted survey activity needs to be undertaken to map the distributions of all the species. The state harbours a significant population of western tragopan, and thus has an important role to play in its conservation. The main populations of this species need to be identified and protected from threats of habitat loss and hunting. Creating public awareness about Galliformes by establishing state of the art display facilities (zoos and aviaries) could be one of the main tasks of the Department of Wildlife Protection, Jammu & Kashmir Government in the coming years. The Department of Wildlife Protection could also join in the conservation breeding efforts of the western tragopan initiated by the Department of Forest and Wildlife, Himachal Pradesh and the Central Zoo Authority (Kaul and Srivastava, 2007).

Uttarakhand

Uttarakhand is India's youngest mountain state, carved out of the erstwhile Uttar Pradesh in the year 2000. The state has diverse habitats ranging from the snow bound peaks of the Himalayas, the highest being Nanda Devi (7,817 m), to the sub-tropical terai region, that contributes to the diversity of flora and fauna of the state. As its commitment to conservation of this valuable natural resource, the state has set aside over 12% of its geographical area (26.6% of its forest area) as National Parks (NP) and Wildlife Sanctuaries.

Uttarakhand is rich in avifauna. Of the 1,248 species of birds reported from India, over 621 have been reported from Uttarakhand. Of these, Galliformes have the pride of place. In the present conservation scenario that is essentially "large mammal" centric, little information is available on the actual conservation status of Galliformes. Most of the information is secondary and often based on anecdotal evidences. However, an attempt is being made to make an assessment of the status of these species in the state.

Status and distribution of Galliformes in Uttarakhand:

1. Red junglefowl: In Uttarakhand, the sal forests of the terai are the main stronghold of the red junglefowl. While commonly seen in Corbett NP, Rajaji NP, and Sona Nadi WS, it is also fairly abundant in other forest tracts of the terai region in the state. Detail information about their population status in the state is lacking. Towards the forest edges, the species can be seen in close proximity to human habitations, which has raised concerns about their interbreeding with domesticated varieties. The Wildlife Institute of India (WII), Dehradun has initiated a project to evaluate the genetic diversity, and other ecological and behavioural aspects of the species in Uttarakhand.

2. Indian peafowl: Although no population estimates are available, the species is abundant in the terai region of the state. It is seen in good numbers in PAs such as the Corbett NP and Rajaji NP. The Jhirna-Kalagarh road is one of the best places to see large number of these birds during an early morning or late evening drive through the forests. It is also seen in good numbers on forest edges close to human habitations. Due to its religious connotations, the species is generally tolerated, despite occasional complaints about damage to crops.

3. Himalayan monal: The Himalayan monal is the State Bird of Uttarakhand and it has a significant presence in the Nanda Devi NP, Valley of Flowers NP, Govind NP, Gangotri NP, Kedarnath WS, Askot WS, and Govind WS. It is also reported from the Chakrata, Badrinath, Bageshwar, and Pithoragarh Forest Divisons. Kedarnath WS in particular, has several excellent locations for sighting of this beautiful bird. Shokharkah, near Tungnath is one such site where patient waiting is usually rewarded by sightings of up to 20-30 individuals in one evening. Encounter rate for monal in this area ranges from 0.7 to 1.2/ km walk and the density estimate ranges from 5 to 20 pairs/km². Madhmaheshwar in Kedarnth WS, and nearby areas are also good for sightings of monal. In Nanda Devi NP, the encounter rate for monal range from 0.75 to 2.28/km walk.

4. Kalij pheasant: Of the 5 sub-species of kalij distributed along the Indian mountain states, Uttarakhand is home to the white crested kalij *(L.l.hamiltonii).* The species exhibits a wide altitudinal distribution, from about 200 to 2,800 m. Thus, in several places it is found with the red junglefowl. The species is common in the temperate forest areas of the state, with Kedarnath WS, Binsar WS, Govind NP & WS, and Askot WS being some of the prominent areas bearing good populations. The Mandal-Chopta road in Kedarnath WS is one of the best places to sight this bird. Abundance estimates for kalij in this area are 8.6 birds/km walk, with a density of 16-17 birds/km². It is also sighted on the foothills and hill sides of the Corbett Tiger Reserve.

5. Cheer pheasant: The cheer pheasant generally occurs in small and isolated populations. In Uttarakhand, the species is rare, reported infrequently from open grassy and scrubby tracts on steep hillsides between 1,500 to 2,500 m. Areas near Lansdowne in Pauri Garhwal district report occasional sightings of the bird. Other areas from where the species has been reported include parts of Nanda Devi Biosphere Reserve, Kedarnath WS, Govind NP & WS, and Tehri and Uttarkashi Forest Divisions. The species continues to be rare, and very limited information is available on its status.

6. Koklass pheasant: This species occurs in well-forested tracts between 2500-3300 m in temperate forests. In Uttarakhand, the Chopta-Mandal road stretch, and the forest tracts around Tungnath are well known

locations to spot the bird. Estimated koklass density of 5 birds/km² in Shokharkh area in Kedarnath WS has been recorded. Other notable areas for the species include parts of Uttarkashi, Tehri and Bageshwar Forest Divisions, and Binsar, Govind WS & NP. In Dibrugeta, Nanda Devi NP, the abundance of koklass was recorded at 3 calling males/ station during spring 2003.

Fig 6.20 A koklass pheasant from '*A Hand-book to the Game-birds*'Vol. I and II, by W.R. Ogilvie-Grant, 1897.

7. Satyr trogopan: The main stronghold of the Satyr trogopan is the eastern Himalayas. However, the species is also reported from parts of Uttarakhand. Sightings are extremely rare and have been confined to some reports from the Pithoragarh, Almora, and Bageshwar Forest Divisions, and the Askot WS.

The landscape in Kumaon Himalaya is a mixture of oak forests, chir pine forest, agricultural fields and human habitations. Landuse practices in past have led to many fold increase in coverage of chir pine forest at the expense of oak forests. Wildlife managers should take steps to halt further degradation of oak patches to ensure survival of kalij, koklass, and other associated faunal and floral elements. The remnant oak patches, highly fragmented, are under tremendous lopping and grazing pressure. The 2 Protected Areas - Binsar and Askot Wildlife Sanctuaries - are not free from grazing, lopping, cutting, and forest fires. In such conditions, both species will be vulnerable due to their requirements for undisturbed habitat conditions. Many of the pheasant species occur equally outside the PA network as within. Firstly, the Protected Area coverage needs to be increased substantially in Kumaon Himalaya (Shah Hussain et al., 1997). Secondly, the managers need to evolve guidelines for regulation of grazing and lopping for protected area as well as outside protected areas, which should be implemented rigorously. Thirdly, the Wildlife Protection Act (1972) should be strictly enforced to curb widespread poaching of all wildlife species. Despite many of the pheasant species finding protection under the provisions of the Wildlife (Protection) Act 1972, their feathers attract many buyers.

One of the major issues for the conservation of galliforms, especially pheasants, is that the various developmental activities across the State attract a large number of migrant labour, who live virtually unsupervised in many interior areas. They are known to occasionally trap these birds for food (as also in Himachal Pradesh), also occasionally for the crest and feathers. The monal crest is proudly worn as traditional attire in some parts of Himachal Pradesh and commands a fancy price. The Forest Department lacks financial and technical resources to implement scientific species management and habitat recovery programmes. The remote and rugged terrain, along with a lack of trained and motivated staff to implement management actions are the constraints for better management in the State. The provision of proper field gear, and more number of patrol cabins will increase the efficiency of patrolling (Sinha and Chandola, 2007).

West Bengal

The State of West Bengal extends from the Himalayas in its north to the Bay of Bengal in its south, and owing to this varying elevation range, it supports a rich biodiversity. The Darjeeling hills form a part of the eastern Himalaya and include Sandakphu—the highest peak of the State. The narrow Terai region separates this region from the plains, which in turn transition into the Ganges delta towards the south. The Rarh region intervenes between the Ganges delta in the east and the western plateau and high lands. A small coastal region is on the extreme south, while the Sundarbans mangrove forests form a remarkable geographical landmark at the Ganges delta. Forests make up 14% of the geographical area of West Bengal; protected forest cover is 4% of the state area. About 19 species of Galliformes occur in West Bengal, and the state has taken several management steps to conserve these birds. Almost all the Galliformes in the state have been protected by declaring their habitat as Protected Areas (PA), however, larger populations of Galliformes still occur outside the PAs, which is a major challenge in conservation of these birds in the state.

Status and distribution of galliformes

1. Red junglefowl is one of the most well known pheasants of West Bengal as it is the most widespread, occurring in the foothill forests of the Duars and Terai, in the Sal forests of western hilly lateritic tracts, and also in the mangroves of Sunderbans. Though this species is widely distributed in the State, so far there has not been any detailed survey to assess the status of this species.

2. Indian peafowl is common in many National Parks (NP) and Wildlife Sanctuaries (WS) in the Duars and Terai in northern West Bengal - in areas such as the Mahananda WS, Garumara NP, Jaldapara WS, and Buxa Tiger Reserve (TR). Small populations still survive in the forests of Purulia and Bankura districts. There is a small population surviving near Bandel, that probably originating from an escaped population in Rajhat, Hooghly district. The green peafowl was reported from Jalpaiguri district in 1957, but it was not sighted again in West Bengal, and probably has become locally extinct there.

3. Kalij pheasant is commonly seen in the undisturbed forests in Darjeeling hills during dawn or dusk. It is also common in the Singhalila NP and Neora Valley NP.

4. Blood pheasant is found in the high altitude (>3200m) areas such as Singhalila NP near Sandakphu.

5. Satyr tragopan inhabits steep hillsides mainly between 2,600–3,800 m. It prefers moist oak and rhododendron forests with dense undergrowth and bamboo clumps. In West Bengal, it is found in the

Singhalila NP and Neora Valley NP. It is a 'Near Threatened' species and needs protection for its long-term survival.

Fig 6.21 A male blood pheasant from 'A Hand-book to the Game-birds' Vol. I and II, by W.R. Ogilvie-Grant, 1897.

Excessive biotic pressures on the forests have led to the degradation as well as reduction in numbers of wild fauna in the state. Grasslands are also subjected to intense pressures from encroachment for settlements and cultivation, over-grazing by domestic livestock, grass harvesting, fire, and tree plantation. Galliformes are also poached in some areas by local people for meat. Development activities, road and railway construction have also led to habitat loss and degradation in some areas.

The state has 4,031.km² of forests, under PA network which is 34% of the state's total forest area and 4.54% of the total geographical area. The

PA network includes 1,055 km² that are covered under 15 WSs, 1,693 km² that are covered under the 5 NPs, and the rest occurs in the buffer zones of the 2 Tiger Reserves, and 1 Biosphere Reserve. These PAs and the Forest Divisions afford protection to several species of Galliformes in the State. West Bengal has been a pioneer in the concept of Joint Forest Management (JFM) for protection of forest resources. This has lately been modified for PAs to be known as Joint Protected Area Management (JPAM). The concept is being tested and modified constantly in the eco-development activities around the PAs, which will ultimately help in the conservation of wildlife, including the Galliformes (Mookherjee, 2007).

Sikkim

One of the smallest and least populous states in India, Sikkim is located between the Himalayan kingdoms of Nepal in the west (the Singalila mountain chain culminating in the mighty Khangchendzonga range at over 8,598m, the world's third highest peak) and Bhutan in the east (Chola Range). Sikkim is bounded by Darjeeling district of West Bengal to the south and a stretch of trans-Himalayan Tibetan plateau in the north. The State is a part of the Himalaya Biodiversity Hotspot.

Forestry is the major landuse in the state and over 80% of the total geographical area of the state under the administrative control of the Department of Forests, Environment and Wildlife management (DFEWL), and 50% of the geographical area is forested. There is 1 high altitude National Park (Khangchendzonga NP) cum Biosphere Reserve, and 7 Wildlife Sanctuaries (WS), which together constitute 42.6% of the total geographical area. A Cold Desert PA for trans-border ungulates such as the Tibetan argali, Tibetan wild ass, and Tibetan gazelle has been proposed in the north (Tso Lhamo Cold Desert Conservation Reserve). The addition of the proposed Protected Areas (PA) in the state would further strengthen the status of Sikkim as the state with the best PA network in the country.

The blood pheasant *Ithaginis cruentus* is the state bird of Sikkim. Satyr tragopan, Himalayan monal and kaleej are also found. As many as 16 species of Galliformes were recorded from this small portion of the Himalayas, ranging from the Tibetan snowcock *Tetraogallus tibetanus* on the Tibetan plateau, to the Indian peafowl *Pavo cristatus* and red junglefowl *Gallus gallus* in the sub-tropical south. Some blue peafowl

Pavo cristatus from Punjab were released into the lowland forests over two decades ago by the DFEWL to augment the reportedly existing population straying across from the West Bengal border. In fact, in the last few years, there have been instances of complaints of crop-raiding by the peafowl in villages around the Kitam Forest. Interestingly, grey peacock pheasants *Polyplectron bicalcaratum* were reported as recently as 1999 from Kewzing, South Sikkim and the recently declared Pangolakha WS in East Sikkim.

The DFEWL carries out regular patrolling for monitoring of habitat to control wildlife offences inside the PAs. However, effective monitoring of Galliformes populations outside PAs is lacking. With newer areas under the PA network, many more management initiatives are being proposed such as intensive scientific research, strengthening of EDC (Ecodevelopment Committees) network, training of forest staff, creation of alternative livelihoods for people dependent on forest resources, and preparation of detailed management plans for the PAs.

Galliformes such as Himalayan monal and Satyr tragopan are also covered under an *ex situ* management programme in the Himalayan Zoological Park at Bulbuley, Gangtok. Two more *ex situ* conservation facilities (a bird park and a butterfly park) are planned (Lachungpa and Bhutia, 2007).

Arunachal Pradesh

The eastern Himalayan mountain chain in northeast India is precious for its rich biological diversity. Arunachal Pradesh represents the easternmost section of the eastern Himalaya and northern section of northeast hills. The State is a part of one of the Global Biodiversity Hotspots namely – Himalaya Biodiversity Hotspot. The state has the largest geographical area among northeast Indian states, due to which large populations of many species occur. It has an altitudinal range from plains (50m) to high mountains (>7,000 m) giving rise to habitats suitable for a whole spectrum of biological elements. The state is known for its avifaunal richness, with more than 700 species of birds. The state forms a large part of the Eastern Himalaya Endemic Bird Area.

Arunachal Pradesh is largely forested, with 13 Protected Areas (PAs) covering 12.8 % of its geographical area.

Fig 6.22 A male Sclater's monal by J G Keulemans in 'Proceedings of the Zoological Society of London', 1870.

The state is home to at least 10 species of pheasants. Of these, Blyth's tragopan *Tragopan blythii* and Sclater's monal *Lophophorus sclateri* are globally threatened. Temminck's tragopan *Tragopan temminckii,* and possibly Tibetan eared pheasant *Crossoptilon harmani* and Mishmi blood Pheasant *Ithaginis cruentus kuseri,* also occur in this State. The State has large forested areas that are still relatively free from biotic and other disturbances (e.g. PAs such as Namdapha NP, Mouling NP, and Dibang WS), due to their remoteness; though large-scale hunting and trapping are responsible for the scarceness of pheasants.

Many areas outside PA network, especially remotely located forested areas in the hills and mountains, hold great potential for a large number of species including the pheasants. Except for a few large PAs (e.g. Dibang Valley WS, Namdpha NP, and Pakke WS) other PAs are small in area, and consequently cover only a very small portion of distributional range of the pheasants. The location of the PAs is also not uniform from east to west, and north to south. For example, there is no PA in the high altitudes north-west of the State. In a state such as Arunachal Pradesh, where large rivers provide dispersal barriers to fauna, there are likely to be distinct populations across the barriers, and therefore any meaningful conservation strategy has to take into consideration the conservation of this variability. This can be achieved if conservation efforts go beyond PA boundaries.

Birdlife International and Bombay Natural History Society (BNHS) have prioritised a network of 'high conservation value bird areas' under its Important Bird Area Programme (IBA). Many of the IBAs are PAs and some of them are Reserved Forests. The limitation of IBA approach is that the IBAs themselves do not have legal status. Therefore protection of such IBAs which are not PAs or Reserved Forests is quite a challenging task. Also, local people should be involved in wildlife conservation and tourism industry; and the eco-tourism activities in the state need to be strengthened (Singh and Singh, 2007).

Assam, Meghalaya, and Tripura

The states of Assam, Meghalaya, and Tripura in northeast India are a part of the Indo-Burma biodiversity hotspot. These states have high habitat diversity, and also a high Galliformes diversity, with 19 species out of a total of 50 species recorded in India. This includes 13 partridges and 6 pheasants. 5 of these are threatened while 3 are range-restricted.

Although their general range has remained almost the same, the expansion of human habitation, the destruction of habitats by agriculture (slash-and-burn shifting cultivation, known as *jhum*), logging, and hunting have resulted in a sharp decline in abundance as well as causing fragmentation. It is worth noting here that slash and burn shifting cultivation was a sustainable method of cultivation among the tribal people of northeast India and some Southeast Asian countries. In a small cleared patch, more than 100 species of edible plants (grains, legumes, tubers, spices, vegetables and medicinal plants) were cultivated. But the increase in tribal population and the simultaneous shrinking of forest land due to intensive agriculture, plantations and urbanization has rendered this method of farming unsustainable.

Although many Galliformes occur within Protected Areas (PA), the enforcement of wildlife laws is inadequate. For the conservation of

Galliformes, developing a long-term strategy in the northeast region of India is of high importance – involving a comprehensive review of the distribution and current status of Galliformes - due to the rapid loss of habitat, increased hunting and fragmentation resulting in small isolated populations.

The states of Assam, Meghalaya, and Tripura contain a major plain in the form of the Brahmaputra valley. There is significant habitat diversity in the region resulting from the large altitudinal ranges (50m to over 2,000m). The main forest types are tropical wet evergreen rainforest, tropical semi-evergreen forest, tropical moist deciduous forest, subtropical broadleaf and coniferous forest, wet savanna grassland, and swamp forest. Since 1985, field studies have been carried out in different parts of northeast India to determine the distribution and status of different species of wildlife, including the Galliformes.

Distribution and Status of Galliformes

1. Blyth's tragopan *Tragopan blythii:* Occurs mainly in subtropical and temperate broadleaf forests in a small area of Assam. In Assam, it has been recorded from eastern areas of Barail Range in North Cachar Hills district. In the Barail Range, it occurs above 1,800 m. It also occurs in the Mt Paona and Laike area of Assam. The historical records from the area include sightings in 1898 and 1899. The current status in the area is unknown as it is a highly disturbed area. However, a small population still occurs as one was shot by a hunter near Laike area in Assam in 2001.

2. Kalij pheasant *Lophura leucomelanos:* The subspecies *L. l. lathami* (black-breasted kalij) is found in the region, with a single record of another sub-species *L. l. moffitii.* A popular game bird that is much sought after by hunters, it is resident and still very common in places. It is more abundant in the south bank areas of the Brahmaputra River where it occurs at 30 m in Karimganj district of Assam, to above 1,900 m in the Barail range. Sightings are frequent in the mornings and evenings. It occupies forested foothills, hills, and the edges of the plains including cultivation near forests. So far it has been recorded in at least 30 PAs and numerous RFs across these states.

3. Red junglefowl *Gallus gallus:* A very common species, also popular as game bird and much sought after by hunters. Found all over the forested

foothills, hills, and plains. It occurs from 30 m in Tripura, and Karimganj district of Assam, to below 1,000 m (rarely up to 2,000 m). Also occupies edges of cultivation near forest, scrub jungle, and wooded tea gardens. It is mostly seen in small groups, often in mixed flocks with kalij pheasant. It has been recorded in at least 30 PAs and numerous RFs across these states.

4. Grey peacock-pheasant *Polyplectron bicalcaratum*: A common and widespread resident, it occurs from 30m elevation in Karimganj district of Assam, to 1,200m in the hills and mountains. Sightings are rare - the birds are more often heard than seen. It has been recorded in more than 18 PAs and numerous RFs across the States. This species is sought after by hunters. It occurs in evergreen as well as deciduous forests. Mostly seen single or in pairs - it is usually not seen at the edges of cultivation near forest and wooded tea gardens.

5. Indian peafowl *Pavo cristatus*: Assam is the easternmost limit of its distribution. It has disappeared from many parts of its former range, and now occurs mainly in western Assam. It has already become extinct in Meghalaya where it was not uncommon in Garo Hills. In Assam, the range is restricted to the *Duars*, from Kokrajhar to Darrang districts, through the northern parts of Bongaigaon, Barpeta, Nalbari, and Kamrup districts (in the reorganised districts of Kokrajhar, Chirang, Baksa and Udalguri). PAs where the species is found are Manas NP and Barnadi WS. Outside PAs, it has become scarce and sightings are not frequent. The easternmost limit of distribution is Rowta RF (92°21'E), which is also now its easternmost world distribution. Unlike in northern India, they are not found in the countryside or in semi-feral state in the region. The species is hunted for its meat.

6. Green peafowl *Pavo muticus*: The subspecies *P.m. spicifer* occurs in India between 50m and 1,000m elevation. It is a very scarce resident with no recent report from these states. In Assam, it was formerly widespread to the south of the Brahmaputra River, from Nagaon to Cachar district. However, there have been no recent reports. The last reports were of a few stragglers from Barak RF (Bhuban Hills) of Cachar district in the 1970s. One specimen was reportedly shot by a hunter in Jirikinding area of Karbi Anglong in the early 1970s. The species has

become locally extinct in most of its range in India mainly due to hunting for its meat.

Several PAs have been created to protect the Galliformes. Galliformes occur within all the 33 NPs and WSs in these states. The degree of protection this provides to different species is highly variable. For example, for the swamp francolin in Assam, more than 65% of its range is within the PA system, whereas for Blyth's tragopan, it is 0%. Except for some PAs, such as Kaziranga NP and Orang NP in Assam, the enforcement of protection measures is inadequate.

The creation of more PAs is strongly recommended. Some high priority areas are: Assam (Kobo *chapori*, Barail area near Laike-Hemplopet) and Meghalaya (Narpuh). Extension of existing PAs such as Balpakram Tiger Reserve, Siju WS, and Nongkhyllem WS in Meghalaya, and Barnadi WS in Assam, is also required. The people's initiative shown by the villagers of Ghosu and Khonoma needs to be replicated by other northeastern states, where bulk of the land belongs to the indigenous community. Some high priority areas for Community Reserves are: Meghalaya (sacred forests near Cherrapunjee and Mawphlang) and in Tripura (near Jampui tlang).

The following measures should be undertaken:
• Conservation education among locals, including the hill tribes of remote areas with active involvement of local NGOs.
• Reduce dependency of fringe villagers on forest. One method is large-scale installation of bio-gas, which will greatly reduce pressure on fuel wood.
• Further research on galliform ecology, behavior, and altitudinal movements in different habitats.
• Where grassland burning is necessary, it should be carried out in January or early February.
• It is important to undertake measures that limit the massive human population growth, especially in the fringe areas, when considering conservation planning (Choudhury et al., 2007).

Manipur, Nagaland, and Mizoram

Northeast India is currently a part of 2 Biodiversity Hotspots – Himalaya and Indo-Burma hotspots. Nagaland, Manipur, Mizoram, and Tripura constitute part of Indo-Burma hotspot. 16 species of Galliformes are found in the three states of Manipur, Nagaland, and Mizoram. There are 6 species of partridges, 4 species of quails, and 6 species of pheasants that are present in these 3 states. Out of these 6 pheasants, 3 species are 'threatened' – Blyth's tragopan, Hume's pheasant, and green peafowl.

Recent surveys recorded Blyth's tragopan from Blue Mountain NP in Mizoram; Zunheboto district, Kohima district, Tuensang district, and Phek district – all in Nagaland. This species is the State Bird of Nagaland. In the year 2000, a trekking expedition to the Dzukou valley on the Manipur-Nagaland border, found a pair of Blyth's tragopan killed by a local hunter. In Manipur, it has been recorded in Siroi.

Fig 6.23 Painting of a pair of Mrs. Hume's pheasants from 'A Monograph of the Pheasants' by William Beebe, 1918-22.

Mrs. Hume's pheasant was recently confirmed based on evidences at the Murlen NP, and sighted at the Blue Mountain NP in Mizoram. This is the State Bird of Mizoram. It has been recorded from north of Siroi Hills, and another area south of Churachandpur in the state of Manipur. Also recorded from 3 areas in Nagaland, *i.e.,* from Phek, west of Mount Saramati and from a forested tract in Tuensang. According to the locals of Siroi village, Ukhrul district, Manipur, there could be a small population of Mrs. Hume's pheasant in the Siroi-Kashong hill range, and the last sighting was around 1984. This is also the State Bird of Manipur. A subsequent survey of Hume's pheasant could locate it in many new sites in Nagaland, Manipur, and Mizoram.

During recent surveys, only secondary evidence of green peafowl was obtained from an area near Moreh at the Indo-Myanmar border in Manipur. This species is also reported from the Khonkhan Thana village area in Ukhrul district along the Indo-Myanmar border. Surveys in Mizoram did not note any evidence of this bird, but a recent report (in 2007) has confirmed the presence of this species in Chintupui and Lunglei districts.

Peacock pheasants are sighted in the Damdei village area in Churachandpur district, Manipur. In the last visit to the area, there was evidence of a large number of the pheasant (and hornbills) recently killed by local hunters for its meat, however, the habitat was found to be in good condition, with primary forest cover over a large area. Interestingly, these pheasants reportedly used several rock shelters found in the area. This species was also found in Murlen NP, and Dampa Tiger Reserve in Mizoram. Elsewhere in Mizoram, it was found in Inner Line RFs, Ngengpui WS, Lengteng WS, Rawlbuk, and Lamzawl.

The 2 species common in all the low and middle altitude forests in the 3 states are kalij pheasant and red junglefowl. However, both face some threat in terms of hunting. A kalij pheasant was recovered dead from a local hunter at Tarao Laimanai village (440m; Chandel district, Manipur) in December 2003. Red junglefowl are noticed in good numbers along the Tegnoupal-New Somtal highway, Chandel district, Manipur. Habitat in the area is primary forest mixed with secondary forests. Junglefowl are also noticed along the Sita Lamkhai-Phungyar state highway, Chandel and Ukhrul districts, and along the Singhat-Behiang highway, Churachandpur district, both in Manipur. There is some good forest cover in the Behiang area, south of the Manipur-Myanmar border, having good wildlife populations.

Galliformes are also found outside of PAs in Manipur. Forests outside the PAs in Mizoram hold substantial populations of hill partridge, mountain bamboo partridge, kalij pheasant, and red junglefowl. In Nagaland, bulk of the Galliformes habitat is outside the PAs. Creation of Community

Reserves by several Village Councils has helped conservation to a great extent.

Threats to the galliform species in these 3 states are as follows:

· Habitat fragmentation and degradation, through a continuing decrease in the rotation period of traditional slash and burn agriculture in forested habitats,
· Reclamation of forested areas for development works such as road and dam building, expansion of human settlements,
· Anthropoid interference such as logging, lopping, and minor forest produce collection etc.
· In Manipur, Nagaland, and Mizoram where hunting is a traditional way of life in the hills, the major threat to Galliformes is random hunting for meat and game.

The lack of species-specific efforts targeted for Galliformes conservation is a problem. Although there are PAs protected by law, such as the Yangoupokpi Lokchao, Kailam and Bunning WS and the proposed Siroi NP which are prime galliform habitats, more effort by the government or other organisations to protect and conserve these wildlife habitats is required. Here also, the provision of proper field gear, and better and more number of patrol cabins will increase the efficiency of patrolling. There is, however, some effort by locals to conserve the Siroi wildlife habitat. For the last many years, the Siroi Youth Club, and more recently an organization called Mungleng Vathei Hill Development Organisation based in Siroi village (Ukhrul district, Manipur), have been engaged in conservation and protection of the Siroi-Kashong hill range. There is also some effort by the Tokpa Kabui villagers (Churachandpur district, Manipur) to conserve their forest which is potential habitat for galliform species. This forest is an important catchment of the Loktak lake and the Leimatak river – a secondary tributary of the Barak river. Similarly, there is a laudable effort by Khambi village (Ukhrul district, Manipur) to conserve community forest area which is a possible junglefowl habitat. In Nagaland, the efforts by the Khonoma village council for declaring the forest adjacent to this village as a 'community conserved forest area' for protecting the habitat of Blyth's tragopans, has become quite successful in protecting the tragopan and other associated species (Ghose et al., 2007).

South India

Southern India is endowed with a rich diversity of Galliformes, including 3 pheasants. However, not much information is available with the exception of some surveys and short-term studies on the distribution, status, habitat requirements, diet, and activity pattern of a few galliform species in southern India. A review of literature clearly indicates that there is a need for more research based information (the current status, distribution and habitat requirements) on Galliformes in southern India. Some of the studies conducted on grey junglefowl and blue peafowl have been described here (Sathyanarayana, 2007).

Grey Junglefowl

The endemic grey junglefowl is confined to peninsular India. Bird surveys in 1990s in Kerala indicated that the population status of grey junglefowl was lower than expected. The occurrence of a small population of grey junglefowl in the northern Kerala near the Cannanore coast indicates that in the past the species had been widespread in Kerala from the coast to the high hills. Random observations over the last 15 years in different parts of the Western Ghats show that grey junglefowl, along with peafowl, red spurfowl, francolins and quail, in this part of India has been declining swiftly, due to poaching and habitat degradation. Investigations on the abundance, group sizes, and sex ratios of grey junglefowl revealed variation between areas and habitats. On Mundanthurai Plateau, the grey junglefowl density estimates ranged from 1.67 to 34.42 groups/km^2 (group size range 1.0 to 1.6) with a mean density of 19.78 groups/km^2 (mean group size 1.3), and the male:female ratio of 1:1.2 (in 2006), and in Theni Forests, Tamil Nadu, it was estimated at 37.03 $\pm$ 2.81 birds/km^2 (in 2007).

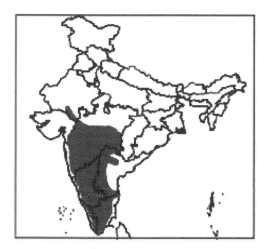

Fig. 6.24 A map showing the distribution of grey junglefowl in India, 2007.

Ecological information on grey junglefowl is limited to some habitat use studies. These include investigations in Bori Wildlife Sanctuary in Madhya Pradesh, Theni in Tamil Nadu, and Mundathurai Plateau in Tamil Nadu. Seven tree species (of 8-20 m height) are preferred for roosting by the grey junglefowl at the Grizzled Giant Squirrel Sanctuary in Tamil Nadu. A study from Anaikatty Hills, Tamil Nadu found that the grey junglefowl used the mixed dry deciduous forest rather than the scrub forest; and use areas that had low to moderate canopy, moderate to high scrub cover, interspersed with areas of low or no grass cover at Mundathurai Plateau, Tamil Nadu. Grey junglefowl is an ominivorous species and food items such as grasshoppers, other insects (beetle, black ant, red ant, etc), exoskeleton of millipedes, egg shells, and undigested seeds of *Capsicum* sp, *Cynodon* sp, *Ziziphus oenoplia,* and *Lantana camara* have been recovered from its fecal matter collected from Theni, Tamil Nadu.

Blue Peafowl

A study has found that, in Scrub jungle and Dry deciduous Forests of Mudumalai Wildlife Sanctuary, Tamilnadu, blue peafowl roost in 8 tree species which includes *Acacia sundra, Cordia obliqua, Bombax malabaricum,* and *Zizphus jujuba.*

The sex ratio of adult male and female has been found to be 1:0.76 at Ketchilapuram Village, Tuticorin District, Tamil Nadu from July 1998 and January 1999. Scientists have also reported seasonal variation in group composition and grouping patterns of Indian peafowl in 3 different seasons. Habitat destruction and poaching for meat and egg were the real threats to the peafowl in this area. Despite the crop damage caused by peafowl, the villagers of Ketchilapuram tolerate the presence of the peafowl due to their religious sentiment. Scientists have done studies on bird pest management with special reference to Indian peafowl in Tamilnadu. In order to protect their crops, the farmers in the villages of Viralimalai Panchayat Union, Pudukottai district, Tamilnadu, used audio tapes, scare-dogs, and crackers to scare the peafowl. The use of reflective ribbon (a polypropylene metallic shining ribbon, with red one side and silver white on other side) was found to be very effective in preventing the peafowl from raiding crops from the paddy fields and other food crops. 8 species of Galliformes occur outside the Protected Area network in Tamil Nadu. There are still gaps in knowledge on the distribution and ecology of galliform species.

Making Conservation a Participatory Process

At the 11th Conference of the Parties (COP 11) to the Convention on Biological Diversity in Hyderabad, India in October 2012, the need was reiterated for increasing funding of projects that help in conserving biodiversity. India's Minister for Environment and Forests declared at the conference that funding for biodiversity conservation will be increased in India in the coming years. Our natural heritage is our national treasure, and economic pressure from urbanization – i.e. mining, manufacturing industry, power generation, urban housing, and transportation sector - should not be allowed to overwhelm wildlife conservation objectives. The areas that require special attention are –

- Improving forest management infrastructure (particularly pressing needs are – revision of the colonial-era working plans to shift focus from forest exploitation to wildlife preservation, and equipping the state Forest Departments with GIS mapping facility),

- Identifying and protecting wildlife corridors for facilitating migration and dispersal, and including more areas in the protected area network for conserving landscapes and species assemblages,
- Regulating trade in endangered flora and fauna – including timber, food, medicine, and pet trade,
- Providing low-cost eco-tourism facilities in protected areas,
- Building educational museums, libraries, and natural history societies in urban areas for the general public; also, installing wildlife sculptures in public spaces, and wildlife art exhibitions,
- Funding the identification, management, and monitoring of 'Green Belts' in urban areas through local municipal governments,
- Taking steps to prevent global warming, and making cities sustainable; redevelopment instead of expansion,
- Increasing the agricultural area under the Integrated Pest Management system of crop protection from the current 2%,
- At the international level, contributing to the identification, assessment, and monitoring of World Heritage Sites, to conserve wildlife and cultural heritage for the future generations.

Improving the level of public-awareness about wildlife conservation, and rigorous efforts by the state governments to systematically identify conservation problems in its protected areas network and other wildlife areas will help to deal with the issues efficiently. The role of NGOs cannot be emphasized enough, and they can provide the much-needed platforms for public participation in protection of the national heritage - by adopting a sustainable lifestyle, and increased willingness to support and donate funds for projects. 'Benefit-sharing' and 'Corporate Social Responsibility' are the new buzz words that will encourage and morally bind the corporate and industrial houses to support biodiversity conservation. There is also a need for increasing the interface between the governments, the scientific, educational, and social NGOs, and the public, so that issues can be highlighted and streamlined for effective solutions. This is applicable to all the pheasant-range countries in Asia and Africa.

Bibliography

1. Ahmed, Abrar (1997), Live Bird Trade in Northern India, New Delhi: TRAFFIC-India, WWF. URL: http://www.traffic.org/birds/
2. Aiyadurai, A. (2011), Wildlife hunting and conservation in Northeast India: a need for an interdisciplinary understanding. *International Journal of Galliformes Conservation*, 2: 61-73. URL: http://www.pheasant.org.uk/uploads/Pages61-73_Aiyadurai.pdf
3. Ali, Salim (1979), *Bird Study in India – Its History and Its Importance*, New Delhi: Indian Council for Cultural Relations. Source – Archive.org. URL: http://archive.org/details/AzadMemorialBirds
4. All India Network Project on Agricultural Ornithiology (2013), Eco-friendly Management of Depredatory Birds, Hyderabad. URL: http://www.ainpao.org/techs.html
5. Althaus, T., Bingaman Lackey, L., Fisken, F. A. & Morgan, D. (2010) The role of international studbooks in conservation breeding programmes. In: *Building a Future for Wildlife: Zoos and Aquariums Committed to Biodiversity Conservation* (ed. by Dick, G. & Gusset, M.), pp. 49–52. Gland: WAZA Executive Office.
6. Anonymous (1998), Book of Odes, *University of Virginia Library,* USA. URL: http://xtf.lib.virginia.edu/xtf/view?docId=Chinese/uvaGenText/tei/shi_jing/AnoShih.xml;chunk.id=AnoShih.1;toc.depth=1;toc.id=AnoShih.1;brand=default
7. Anonymous, Captive Breeding and Reintroduction. JRank Science & Philosophy website. URL: http://science.jrank.org/pages/1191/Captive-Breeding-Reintroduction-Reintroduction.html#ixzz0WQcOpfUP
8. Assink, Han (1995), "Captive Breeding Initiatives for the Vietnamese *Lophura* Species". In newsletter of the WPA/BirdLife/IUCN SSC, *Tragopan* issue 3, 1995. URL: http://www.galliformes-sg.org/pheamain.html
9. Athreya, R. (2006), Eaglenest Biodiversity Project (2003 – 2006): Conservation resources for Eaglenest wildlife sanctuary. Kaati Trust, Pune. URL: http://ncra.tifr.res.in/~rathreya/Eaglenest/ebp1reportW.pdf
10. Bagchi, S. & Mishra, C. (2004) Living with large carnivores - snow leopard predation on livestock in the Spiti - Trans Himalaya. NCF Technical Report No. 11, Mysore, India: Nature Conservation Foundation. URL: *http://www.conservation.in/publication.php?type=Technical+Report&title=55*

11. Baker, E.C. Stuart (1922), *The Fauna of British India including Ceylon and Burma – Birds, Vol 1.*, Publ. London: Taylor and Francis. Source- Archive.org. URL: http://archive.org/details/faunaofbritishin01bake

12. Beebe, William (1927), *Pheasant Jungles*, New York City: Blue Ribbon Books Inc. Source- Archive.org. URL: http://archive.org/details/pheasantjungles031705mbp

13. Beebe, William (1922), *A Monograph of the Pheasants*, Vol. 3. New York Zoological Society and H.F. & G. Witherby Publ., London, England. URL: http://biodiversitylibrary.org/bibliography/50777#/summary

14. Bhadouria, B S, Mathur, V B, Kaul, Rahul (2012), Monitoring of organochlorine pesticides in and around Keoladeo National Park, Bharatpur, Rajasthan, India, *Environmental Monitoring and Assessment*, September 2012, Volume 184, Issue 9, pp 5295-5300.

15. BirdLife International (2012) IUCN Red List for birds. Downloaded from http://www.birdlife.org on 24/12/2012.

16. BirdLife International (2001), Threatened Birds of Asia: The BirdLife International Red Data Book. Cambridge: Birdlife International. URL: http://www.birdlife.org/datazone/info/spcasrdb

17. BirdLife International (2003), Launch of "Saving Asia's Threatened Birds", *The Babbler*, December 2003, No. 8. Birdlife International in Indochina. URL: http://www.birdlifeindochina.org/birdlife/report_pdfs/babbler8.pdf

18. BirdLife International (2014) Species factsheet: *Afropavo congensis*. Downloaded from http://www.birdlife.org on 16/01/2014.

19. Bocxstaele, R. Van (1992), "Studbooks and breeding registers for Pheasants" in David Jenkins (ed.) Proceedings of the 5th Intl. Symposium on Pheasants in Asia, Lahore, Pakistan. WPA.

20. Carroll, John P. (2011), Summing up of Symposium, *G@llinformed*, Newsletter of the IUCN-SSC/WPA Galliformes Specialist Group, UK. URL: http://www.galliformes-sg.org/gallnewspdf/G@llinformed4.pdf

21. Central Zoo Authority, Concept paper on In-situ ex-situ linkage -Conservation Breeding of Endangered Wild Animal Species in India. URL: http://www.cza.nic.in/Concept%20paper%20on%20In%20&%20ex-situ%20linkage.pdf

22. Chang-qing, Ding (1995), "Fieldwork Conditions while Studying Cabot's Tragopan". In newsletter of the WPA/BirdLife/IUCN SSC, *Tragopan* issue 2, 1995. URL: http://www.galliformes-sg.org/pheamain.html

23. Chang-qing, Ding (1995), "The Mating System and Breeding Ecology of Cabot's Tragopan *Tragopan caboti"*. In newsletter of the WPA/BirdLife/IUCN SSC, *Tragopan* issue 2, 1995. URL: http://www.galliformes-sg.org/pheamain.html

24. Chauhan, Anita (2008), 'The Pheasants of the World', Blog, New Delhi. URL: www.pheasantsoftheworld.blogspot.com

25. Chauhan, Anita (2012), Green Infrastructure and Guided Nature Walks for Urban Sustainability – A Study for the City of Shimla. Unpublished P.G. Diploma project report, New Delhi: National Law University and WWF-India Center for Environmental Law, New Delhi.

26. Chauhan, Anita (2013), "Pesticides, Farm Ecosystems and the Blue Peafowl", *Mor* – Newsletter of World Pheasant Association –India, March 2013, No. 18: 10-11.

27. Chauhan, Anita (2013), "Great Himalayan National Park: Dealing with People's Rights", *Mor* – Newsletter of World Pheasant Association-India, November 2013, No. 20: 6-7.

28. Chauhan, Kuldeep (2011), "Guchhi collectors clamour for govt control", The Tribune, Chandigarh, India, May 20, 2011. URL: http://www.tribuneindia.com/2011/20110520/main7.htm

29. Chauhan, Pratibha (2013), "Foreign experts to help check leopard attacks", The Tribune, Chandigarh, India, September 24, 2013. URL: http://www.tribuneindia.com/2013/20130925/himachal.htm#2

30. Choudhury, Anwaruddin, Goswami, Anil Kumar and Lahkar, Kulojyoti (2007), "Status, Distribution and Management of Galliformes in Assam, Meghalaya and Tripura". In S. Sathyakumar and K. Sivakumar (Eds.). *Galliformes of India*. ENVIS Bulletin: Wildlife and Protected Areas, Vol. 10 (1). Wildlife Institute of India, Dehradun, India.

31. Choudhury, B.C., Sathyakumar, S. and Sylvia, Christi (2007), "An Assessment of the Current Status of the Indian Peafowl (*Pavo cristatus*) in India based on Questionnaire Surveys". In S. Sathyakumar and K. Sivakumar (Eds.). *Galliformes of India*. ENVIS Bulletin: Wildlife and Protected Areas, Vol. 10 (1). Wildlife Institute of India, Dehradun, India.

32. CITES (2013), 'What is CITES?' *URL: http://www.cites.org/eng/disc/what.php*

33. CITES (2013), COP16 - Proposals for amendment of Appendices I and II. URL: http://www.cites.org/eng/cop/16/prop/index.php

34. Collar, Nigel J. (2009), "Hainan Peacock-Pheasant: Another CR Species for the IUCN RED LIST?" *G@llinformed*, Newsletter of the IUCN-SSc/WPA Galliformes

Specialist Group, UK. URL: http://www.galliformes-sg.org/gallnewspdf/G@llinformed2.pdf

35. Collins, N.M., Sayer, J.A., Whitmore, T.C. (eds.)(1991), *The Conservation Atlas of Tropical Forests of Asia and the Pacific,* London: Macmillan Press Ltd. Reproduced with permission of Palgrave Macmillan. URL: http://www.unep-wcmc.org/the-conservation-atlas-of-tropical-forests-asia_70.html

36. Convention on Biological Diversity (2012), COP 11, Hyderabad. URL: http://www.cbd.int/cop11

37. Corder, John (2003), "Somerset Wildlife Park to help endangered Reeve's Pheasant". In newsletter of the WPA/BirdLife/IUCN SSC, *Tragopan* issue 18, 2003. URL: http://www.galliformes-sg.org/pheamain.html

38. Corder, John (2007), "Conserving the Galliformes of India: Ex situ Conservation Breeding and Re-introduction". In S. Sathyakumar and K. Sivakumar (Eds.). *Galliformes of India.* ENVIS Bulletin: Wildlife and Protected Areas, Vol. 10 (1). Wildlife Institute of India, Dehradun, India.

39. Danielsen, Finn et al (2007), Increasing Conservation Management Action by Involving Local People in Natural Resource Monitoring. *Ambio* Vol. 36, No. 7.

40. Danielsen, F., Burgess, N. D., Jensen, P. M. and Pirhofer-Walzl, K. (2010), Environmental monitoring: the scale and speed of implementation varies according to the degree of people's involvement. *Journal of Applied Ecology*, 47: 1166–1168. URL: http://onlinelibrary.wiley.com/doi/10.1111/j.1365-2664.2010.01874.x/full

41. Danielsen, Finn et al (2013), Linking public participation in scientific research to the indicators and needs of international environmental agreements, *Conservation Letters* 00 (2013) 1–13.

42. Davison, G.W.H. and McGowan, P.J.K. (2009), "Is the Double-banded Argus *Argusianus bipunctatus* a valid species?" *Birding Asia* No. 12, Bulletin of the OBC.

43. Dennis, R. A., E. Meijaard, R. Nasi, and L. Gustafsson (2008). Biodiversity conservation in Southeast Asian timber concessions: a critical evaluation of policy mechanisms and guidelines. *Ecology and Society* **13**(1): 25. URL: http://www.ecologyandsociety.org/vol13/iss1/art25/

44. Dhiman, Sat Pal (2010), "Small Wildlife Populations face high Extinction Risks" in Sanjeeva Pandey (ed.) Wildlife Week Souvenir, Shimla: Wildlife Wing, Himachal Pradesh Forest Department. URL: http://hpforest.gov.in/pdf/2010wildlife.pdf

45. Dhiman, Satpal (2011), "Conservation of cheer pheasants in Himachal Pradesh". Personal Communication.

46. Dowell, S.D., Aeblscher, N.J., & Robertson, P.A. (1992), "Analysing habitat use from radio-tracking data" in D. Jenkins (ed.) Proceedings of the 5th International Symposium on Pheasants in Asia, Lahore, Pakistan. World Pheasant Association, Reading.

47. Draycott, Roger A.H., Bliss, Thomas H., Carroll, John P. and Pock, Karl (2009), Provision of brood-rearing cover on agricultural land to increase survival of wild ring-necked pheasant *Phasianus colchicus* broods at Seefeld Estate, Lower Austria, Austria, *Conservation Evidence* 6, 6-10. URL: http://www.ConservationEvidence.com

48. Eligon, John (2012), "As Pheasants Disappear, Hunters in Iowa Follow", The New York Times, Iowa, December 31, 2012. URL:*http://www.nytimes.com/2013/01/01/us/as-pheasants-disappear-hunters-in-iowa-follow.html?pagewanted=1&_r=1*

49. FairWild Foundation (2010), FairWild Standard Version 2.0, Switzerland: FairWild Foundation. URL: http://www.fairwild.org/documents/

50. Fernandes, Merwyn et al (2007), "Conservation of Red Junglefowl in India-Towards Mapping Abundance and Genetic Diversity". In S. Sathyakumar and K. Sivakumar (Eds.). *Galliformes of India*. ENVIS Bulletin: Wildlife and Protected Areas, Vol. 10 (1). Wildlife Institute of India, Dehradun, India.

51. Fuller, R.A. and Garson, P.J. (eds.). (2000). Pheasants. Status Survey and Conservation Action Plan 2000–2004. WPA/ BirdLife/SSC Pheasant Specialist Group. IUCN, Gland. Switzerland and Cambridge, UK and the World Pheasant Associatiion, Reading, UK. vii + 76 pp.

52. Garson, Peter J. (2007), "Pheasant Conservation in India (1980-2008)". In S. Sathyakumar and K. Sivakumar (Eds.). *Galliformes of India*. ENVIS Bulletin: Wildlife and Protected Areas, Vol. 10 (1). Wildlife Institute of India, Dehradun, India.

53. Gaston A. J., Garson P. J., Hunter M. L. (1983). The Status and Conservation of Forest Wildlife in Himachal Pradesh, Western Himalayas. *Biol. Conserv.* 27: 291-314. URL: http://pdf.usaid.gov/pdf_docs/PNAAQ722.pdf

54. Gaston, A.J. (1982), "Surveys, Census, Monitoring and Research – their role in Pheasant conservation", in C.D.W. Savage and M. W. Ridley *(eds.) Pheasants in Asia, Proceedings of the 2nd International Pheasant Symposium held in Srinagar, Kashmir*. New Delhi: World Pheasant Association.

55. Ghose, Dipankar (2000), "Hume's Pheasant Sightings in Mizoram, India". In newsletter of the WPA/BirdLife/IUCN SSC, *Tragopan* issue 12, 2000. URL: http://www.galliformes-sg.org/pheamain.html

56. Ghose, Dipankar, Lobo, Peter, Rajesh, Salam and Choudhury, Anwaruddin (2007), "Status, Distribution and Management of Galliformes in Manipur,

Nagaland and Mizoram". In S. Sathyakumar and K. Sivakumar (Eds.).
Galliformes of India. ENVIS Bulletin: Wildlife and Protected Areas, Vol. 10 (1).
Wildlife Institute of India, Dehradun, India.

57. Government of India (2009) CITES Annual Report, Ministry of Environment
and Forests, New Delhi. URL:
http://wccb.gov.in/Content/AnnualReports.aspx

58. Government of India, Madhya Pradesh Forest Department (2012). URL:
http://www.mpforest.org/index.html

59. Grewal, Bikram (1993), *Birds of India, Bangladesh, Nepal, Pakistan and Sri
Lanka*, New Delhi and HK: Guidebook Company Limited.

60. Gulati, A. K., Pandey, S., Gupta, Satish, and Gupta, Kanta (2004), *Enchanting
Himachal – A Guide to National Parks and Wildlife Sanctuaries of Himachal
Pradesh*. Publ. by Wildlife Wing, Himachal Pradesh Forest Department.

61. Hance, Jeremy (2009), "New rainforest reserve in Congo benefits bonobos
and locals". Source- Mongabay.com. San Francisco, CA. URL:
http://print.news.mongabay.com/2009/0525-hance_kokolopori.html

62. Hilaluddin, Kaul, R. & Ghose, D. (2005), "Extraction and use of Galliformes by
indigenous ethnic groups in Northeast India". Proceedings of the 3rd
International Galliformes Symposium, World Pheasant Association,
Fordingbridge, UK.

63. Hobley, Mary (1992), Policy, Rights, and Local Forest Management: The Case
of Himachal Pradesh, India. Rural Development Forestry Network, UK.

64. Holling, C.S., editor (1978), *Adaptive Environmental Assessment and
Management*. John Wiley & Sons, New York, New York.

65. Howman, K.C.R. and Singh, S. (1982), "India's Pheasants and their
Conservation", in C.D.W. Savage and M.W. Ridley (eds.) Pheasants in Asia,
Proceedings of the 2nd International Pheasant Symposium held in Srinagar,
Kashmir. New Delhi: World Pheasant Association.

66. Howman, K.C.R. and Sivelle, Charles (1982), "Pheasant Aviculture and its Role
in Conservation", in C.D.W. Savage and M. W. Ridley (eds.) Pheasants in Asia,
Proceedings of the 2nd International Pheasant Symposium held in Srinagar,
Kashmir. New Delhi: World Pheasant Association.

67. Hume, A.O. and Marshall, C.H.T. (1879), *The Game Birds of India Burmah,
and Ceylon*, vol. 1. Source- Archive.org. URL:
http://www.archive.org/details/GameBirdsOfIndia1

68. International Tropical Timber Organisation (2006), International Tropical Timber Agreement, Geneva, Switzerland: UNCTD. URL: http://www.itto.int/itta/

69. IUCN (2013). IUCN Red List of Threatened Species. Version 2013.1. URL: www.iucnredlist.org. Downloaded on 16 August 2013.

70. IUCN (2001), IUCN Red List Categories and Criteria: Version 3.1. IUCN Species Survival Commission. IUCN, Gland, Switzerland and Cambridge, UK. ii + 30 pp.

71. Jayapal, R., Sivakumar, K., Sathyakumar, S. and Mathur V.B. (2007), "Biogeographical Analysis of Galliformes Distribution in India and Pheasants in the Himalayan Protected Areas". In S. Sathyakumar and K. Sivakumar (Eds.). *Galliformes of India*. ENVIS Bulletin: Wildlife and Protected Areas, Vol. 10 (1). Wildlife Institute of India, Dehradun, India.

72. Jenkins, Kurt, Woodward, Andrea and Schreiner, Ed (2003), A Framework for Long-term Ecological Monitoring in Olympic National Park: Prototype for the Coniferous Forest Biome: Federal Government Series - Information and Technology Report - 2003-0006, Publ. by U.S. Fish and Wildlife Service.

73. Johnsgard, P.A. (1999), *The Pheasants of the World - Biology and Natural History*, 2nd Edition, Washington, D.C.: Smithsonian Institution Press.

74. Kalsi, Rajiv S. (2001), "Status and habitat of Cheer pheasant in Himachal Pradesh, India: Results of 1997-1998 surveys". In newsletter of the WPA/BirdLife/IUCN SSC, *Tragopan* issue 13/14, 2001. URL: http://www.galliformes-sg.org/pheamain.html

75. Kalsi, Rajiv S. (2007), "Status, Distribution and Management of Galliformes in Arid and Semi-arid Zones of India". In S. Sathyakumar and K. Sivakumar (Eds.). *Galliformes of India*. ENVIS Bulletin: Wildlife and Protected Areas, Vol. 10 (1). Wildlife Institute of India, Dehradun, India.

76. Kaul, Rahul (2007), "Conservation of Galliformes in the Indian Himalaya". In S. Sathyakumar and K. Sivakumar (Eds.). *Galliformes of India*. ENVIS Bulletin: Wildlife and Protected Areas, Vol. 10 (1). Wildlife Institute of India, Dehradun, India.

77. Kaul, Rahul and Srivastava, A.K. (2007), "Status, Distribution and Conservation of Galliformes in Jammu and Kashmir". In S. Sathyakumar and K. Sivakumar (Eds.). *Galliformes of India*. ENVIS Bulletin: Wildlife and Protected Areas, Vol. 10 (1). Wildlife Institute of India, Dehradun, India.

78. Khaling, Sarala (1997), "Satyr Tragopan in Singhalila National Park, Darjeeling, India". In newsletter of the WPA/BirdLife/IUCN SSC, *Tragopan* issue 6, 1997. URL: http://www.galliformes-sg.org/pheamain.html

79. Kothari, Ashish (2007), *Birds in Our Lives*, Hyderabad, India: Universities Press.

80. Lachungpa , Usha and Bhutia, N.T. (2007), "Status, Distribution and Management of Galliformes in Sikkim". In S. Sathyakumar and K. Sivakumar (Eds.). *Galliformes of India*. ENVIS Bulletin: Wildlife and Protected Areas, Vol. 10 (1). Wildlife Institute of India, Dehradun, India.

81. Leftwich, A.W. (1983), *A Dictionary of Zoology*, 1st Edition, India: CBS Publ. and Distributers.

82. Madge, Steve and McGowan, P.J.K. (2002), *Pheasants, Partridges and Grouse*, London: Christopher Helm, an imprint of A&C Black (Publishers) Ltd.

83. Mahabal, Anil and Tak, P.C. (2002), *Status Survey of Endangered Species*: Status and review of the Western Tragopan *Tragopan melanocephalus* (J. E. Gray) in India: 1-19. (Published by Director, Zoological Survey of India, Kolkata).

84. Maunder, Mike and Byers, Onnie (2004), Technical Guidelines on the Management of *Ex Situ* Populations for Conservation, *Oryx* 38: 342-346. URL: http://www.zooprint.org/zooprintmagazine/2009/july/full_magazine.pdf

85. McGowan, P.J.K. (1994), "Family Phasianidae (Pheasants and Partridges)" in Josep del Hoyo, Andrew Elliott, and Jordi Sargatal (eds.) *Handbook of the Birds of the World*, Volume 2. (New World Vultures to Guineafowl), Barcelona : Lynx Edicions.

86. Mishra, C. (1997) "Livestock grazing and wildlife conservation in the Indian trans-Himalaya: a preliminary survey". NCF Technical Report No. 1, Mysore, India: Nature Conservation Foundation. URL: http://www.ncf-india.org/publication.php?type=Technical+Report&title=45

87. Mookherjee, Kushal (2007), "Status, Distribution and Management of Galliformes in West Bengal". In S. Sathyakumar and K. Sivakumar (Eds.). *Galliformes of India*. ENVIS Bulletin: Wildlife and Protected Areas, Vol. 10 (1). Wildlife Institute of India, Dehradun, India.

88. Mourer-Chauvire, Cecile (1989), A peafowl from the Pliocene of Perpignan, France, *Palaeontology*, Vol. 32, Part 2, pp. 439-446.

89. Neang, Thy (2009), Liquid resin tapping by local people in Phnom Samkos Wildlife Sanctuary, *Cambodian Journal of Natural History*, 2009 (1), 16-25. URL: Http://www.fauna-flora.org/wp-content/uploads/CJNH_2009_entire.pdf

90. O'Brien, Timothy G. and Kinnaird, Margaret F. (1997), " Wildlife Conservation Society Surveys for Bornean Peacock Pheasant". In newsletter of the WPA/BirdLife/IUCN SSC, *Tragopan* issue 6, 1997. URL: http://www.galliformes-sg.org/pheamain.html

91. Pandey, Sanjeeva and Tandon, Vinay (2007), "Status, Distribution and Management of Pheasants in Himachal Pradesh". In S. Sathyakumar and K. Sivakumar (Eds.). *Galliformes of India*. ENVIS Bulletin: Wildlife and Protected Areas, Vol. 10 (1). Wildlife Institute of India, Dehradun, India.

92. Pisharoti, Priyanka Mohan (2008), 'Livelihood changes in response to restrictions on resource extraction from the Great Himalayan National Park', unpublished MS dissertation submitted to Manipal University, Bangalore, India.

93. Rajan, Chandra (1989), *Kalidasa – The Loom of Time*, Penguin Books India, New Delhi.

94. Ramesh, K. et al (2002), "Report on Radio Tracking of Western Tragopan in the Great Himalayan National Park, India". In newsletter of the WPA/BirdLife/IUCN SSC, *Tragopan* issue 16, 2002. URL: http://www.galliformes-sg.org/pheamain.html

95. Rantanen, E. M., Buner, F., Riordan, P., Sotherton, N. and Macdonald, D. W. (2010), Habitat preferences and survival in wildlife reintroductions: an ecological trap in reintroduced grey partridges. *Journal of Applied Ecology*, 47: 1357–1364. URL: http://onlinelibrary.wiley.com/doi/10.1111/j.1365-2664.2010.01867.x/full

96. Rimlinger, David et al (2001), "Surveys for Sclater's Monal in Northwestern Yunnan". In newsletter of the WPA/BirdLife/IUCN SSC, *Tragopan* issue 13/14, 2001. URL: http://www.galliformes-sg.org/pheamain.html

97. Sathyakumar, S. and Sivakumar, K. (Eds.). 2007. Galliformes of India. ENVIS Bulletin: Wildlife and Protected Areas, Vol. 10 (1). Wildlife Institute of India, Dehradun, India. 252.pp. URL: http://wiienvis.nic.in/PublicationDetails.aspx?SubLinkId=129&LinkId=627&Year=2007

98. Sathyakumar, S., Kaul, Rahul and Kalsi, Rajiv (2007), "Techniques for Monitoring Galliformes in India". In S. Sathyakumar and K. Sivakumar (Eds.). *Galliformes of India*. ENVIS Bulletin: Wildlife and Protected Areas, Vol. 10 (1). Wildlife Institute of India, Dehradun, India.

99. Sathyanaryana, M. C. (2007), "Review of Research on Pheasants in Southern India". In S. Sathyakumar and K. Sivakumar (Eds.). *Galliformes of India*. ENVIS Bulletin: Wildlife and Protected Areas, Vol. 10 (1). Wildlife Institute of India, Dehradun, India.

100. Sathyanaryana, M.C. (2007), "Impact of the Indian Peafowl (*Pavo cristatus*) on Agricultural ecosystems". In S. Sathyakumar and K. Sivakumar (Eds.). *Galliformes of India*. ENVIS Bulletin: Wildlife and Protected Areas, Vol. 10 (1). Wildlife Institute of India, Dehradun, India.

101. Setha, Tan and Bunnat, Pich (2002), "The status and distribution of green peafowl *Pavo muticus* in southern Mondulkiri Province, Cambodia". In newsletter of the WPA/BirdLife/IUCN SSC, *Tragopan* issue 16, 2002. URL: http://www.galliformes-sg.org/pheamain.html

102. Severinghaus, Lucia Liu (1996), "Swinhoe's Pheasant in Yushan National Park". In newsletter of the WPA/BirdLife/IUCN SSC, *Tragopan* issue 5, 1996. URL: http://www.galliformes-sg.org/pheamain.html

103. Shah Hussain, M. and Sultana, Aisha (2013), Social Organisation and Sex Ratio among Four Species of Himalayan Pheasants during Breeding and Non Breeding Seasons, *International Journal of Ecology and Development*, Volume 24, Issue No. 1

104. Shah Hussain, M., Khan, Jamal A. and Kaul, Rahul (2001), Aspects of ecology and conservation of Kalij *Lophura leucomelana* and Koklas *Pucrasia macrolopha* in the Kumaon Himalaya, India, *Tropical Ecology* 42(1): 59-68, 2001 ISSN 0564-3295. URL: http://www.tropecol.com/pdf/open/PDF_42_1/42107.pdf

105. Shah Hussain, M., Khan, Jamal A., Ahmad, A. and Kaul, Rahul (1997), Status and conservation of Galliformes in the Kumaon Himalaya, Uttar Pradesh, India, *International Journal of Ecology and Environmental Sciences* 23: 409-417.

106. Singh R. and Singh, S. (1982), "Cheer Pheasants in India", in C.D.W. Savage and M.W. Ridley (eds.) Pheasants in Asia, Proceedings of the 2nd International Pheasant Symposium held in Srinagar, Kashmir. New Delhi: World Pheasant Association.

107. Singh, K.R. and Singh, K.S. (1995), *Pheasants of India – and their Aviculture*. Publ. by Wildlife Institute of India.

108. Singh, Pratap and Singh, K.D. (2007), "Status, Distribution and Management of Pheasants in Arunachal Pradesh". In S. Sathyakumar and K. Sivakumar (Eds.). *Galliformes of India*. ENVIS Bulletin: Wildlife and Protected Areas, Vol. 10 (1). Wildlife Institute of India, Dehradun, India.

109. Singh, Somendra and Tu, Frederikke (2008), A Preliminary Survey for Western Tragopan *Tragopan melanocephalus* in the Daranghati Wildlife Sanctuary, Himachal Pradesh. *Indian Birds* Vol. 4 No. 2. URL: http://www.indianbirds.in/pdfs/Western%20Tragopan.pdf

110. Sinha, Samir and Chandola, S. (2007), "Status, Distribution and Management of Galliformes in Uttarakhand". In S. Sathyakumar and K. Sivakumar (Eds.).

Galliformes of India. ENVIS Bulletin: Wildlife and Protected Areas, Vol. 10 (1). Wildlife Institute of India, Dehradun, India.

111. Snow Leopard Trust (2013), How to Keep Snow Leopards Away From Your Fridge. Snowleopard.org. URL: http://www.snowleopard.org/how-to-keep-snow-leopards-away-from-your-fridge

112. Srivastav, Ajay (2010), Western Tragopan Census. H.P. Forest Department. URL: http://www.scribd.com/doc/37332511/Western-Tragopan-Census

113. The Cornell Lab of Ornithology (2007), Molting, URL: http://www.birds.cornell.edu/AllAboutBirds/studying/feathers/molting/document_view

114. Tucker, Richard (1997), The Historical Development of Human Impacts on Great Himalayan National Park, FREEP-GHNP Research Project, WII, Dehradun.

115. USDA (2013), Conservation Reserve Program, Washington, DC, URL:*http://www.fsa.usda.gov/FSA/webapp?area=home&subject=copr&topic=crp*

116. Vasan, Sudha (1998), 'Political Ecology of Timber Rights in the Western Himalayas', IGES International Workshop on Forest Conservation Strategies for the Asia and Pacific Region, July, 1998, Kanagawa, Japan. URL: http://pub.iges.or.jp/modules/envirolib/upload/1508/attach/complete-report.pdf.

117. Vasan, Sudha (2000), Policy, Practice and Process: Understanding Policy Implementation. Fellows' Weekly Seminar, IIAS, Shimla, India.

118. Watkins, C.W., Barrett, A.M., Smith, R., Paine, J.R. (1997), *Protected Areas Systems Review of the Indo-Malayan Realm,* Ed. John MacKinnon, England: Asian Bureau of Conservation and World Bank. URL: http://www.unep-wcmc.org/protected-areas-systems-review-of-the-indo-malayan-realm_180.html

119. Wich SA, Gaveau D, Abram N, Ancrenaz M, Baccini A, et al. (2012) Understanding the Impacts of Land-Use Policies on a Threatened Species: Is There a Future for the Bornean Orang-utan? PLoS ONE 7(11): e49142. doi:10.1371/journal.pone.0049142. URL: http://www.plosone.org/article/info:doi/10.1371/journal.pone.0049142

120. Wikipedia – the free encyclopedia. Sites for 'Dr. T.C. Jerdon', 'Mr. A.O. Hume', 'CHT Marshall', 'Mr. E. W. Oates', 'Dr. WT Blanford', 'E.C. Stuart Baker', 'Dr. C.W. Beebe', 'Robert Swinhoe', 'Col. Charles Swinhoe', 'P.L. Sclater', 'C. J. Temminck', 'Hon. Walter Rothschild', 'Daniel Giraud Elliot', 'Mr. W. B. Tegetmeier', 'Mr. Frank Finn', 'Mr. W.R. Ogilvie-Grant', 'Jean T.

Delacour', Western Ghats, Argus pheasant, and 'Twitching'. Wikimedia Foundation, Inc. URL: http://www.wikipedia.org/

121. Wildlife Trust of India (2011), Two new site records for rare bird species in Jammu and Kashmir. Source- Wildlife Trust of India, U.P. URL: http://www.wildlifetrustofindia.org/current-news/110610_Two_new_site_records_for_rare_bird_species_in_Jammu_and_Kashmir.html

122. Winarni, Nurul L. (2002), "Studies on the abundance and distribution patterns of great argus *Argusianus argus* in Bukit Barisan Selatan National Park, Sumatra, Indonesia". In newsletter of the WPA/BirdLife/IUCN SSC, *Tragopan* issue 17, 2002. URL: http://www.galliformes-sg.org/pheamain.html

123. World Pheasant Association and IUCN/SSC Re-introduction Specialist Group (eds.) (2009). Guidelines for the Re-introduction of Galliformes for Conservation Purposes. Gland, Switzerland: IUCN and Newcastle-upon-Tyne, UK: World Pheasant Association. 86 pp. URL: http://data.iucn.org/dbtw-wpd/edocs/ssc-op-041.pdf

124. Zacharias, V.J. (1997), "Status of the Grey Junglefowl Gallus sonneratii in Periyar Tiger Reserve, Kerala, S. India". In newsletter of the WPA/BirdLife/IUCN SSC, *Tragopan* issue 7, 1997. URL: http://www.galliformes-sg.org/pheamain.html

125. Zurick, D; Pacheco, J; Shrestha, B; Bajracharya, B (2005) *Atlas of the Himalaya*. Kathmandu: ICIMOD. URL: http://lib.icimod.org/record/7525

126. World Wide Fund for Nature (2013 a), 'List of Terrestrial Ecoregions – Tropical and Sub-tropical Moist Broadleaf Forests – Indo Malayan'. URL: http://wwf.panda.org/about_our_earth/ecoregions/ecoregion_list/

127. World Wide Fund for Nature (2013 b), 'Palm Oil Market and Sustainability in India – WWF Report IND, 2013'. WWF-India Secretariat, New Delhi.

Recommended Websites

1. Wildlife photos and videos on ARKive: http://www.arkive.org/
2. World Pheasant Association WPA: http://www.pheasant.org.uk/
3. International Union for Conservation of Nature IUCN: http://iucn.org/
4. BirdLife International: http://www.birdlife.org/index.html
5. CITES: http://www.cites.org/
6. WAZA: http://www.waza.org/en/site/home
7. ISIS: http://www.isis.org/pages/aboutus_overview.aspx
8. Blog: www.pheasantsoftheworld.blogspot.com
9. Digital Library: www.archive.org
10. UNESCO World Heritage Committee: http://whc.unesco.org/
11. Rainforest Alliance: http://www.rainforest-alliance.org/green-living/marketplace
12. Forest Stewardship Council: https://ic.fsc.org/
13. Sky Island Alliance: http://www.skyislandalliance.org/
14. WWF Maps: http://worldwildlife.org/biomes
15. Salim Ali: http://www.archive.org/stream/AzadMemorialBirds/AliBirdstudy#page/n1/mode/2up
16. Buceros, BNHS, 1999: http://www.archive.org/stream/HistoryOfIndianOrnithology/IndianOrnithology#page/n0/mode/2up
17. *The Malay Archipelago* by Alfred R. Wallace: http://archive.org/stream/malayarchipelago00wall#page/n9/mode/2up
18. ITTO: http://www.forestlandscaperestoration.org/media/uploads/File/ITTOGuidelines.pdf
19. Central Zoo Authority: http://www.cza.nic.in/Final%20Report%20of%20the%20Project%20on%20%20Record%20Keeping%20in%20zoos%20by%20Darjeeling%20Zoo.pdf
20. Royal Society for the Protection of Birds: http://www.rspb.org.uk/supporting/campaigns/sumatra/
21. Harapan Rainforest: http://www.harapanrainforest.org/
22. Saving Asia's Threatened Birds: http://www.birdlife.org/action/science/species/asia_strategy/asia_strategy.html
23. Orientalbirdclub.org: http://orientalbirdimages.org/

24. The Internet Bird Collection - http://ibc.lynxeds.com/
25. Birding: www.Fatbirder.com
26. http://www.ecologyandsociety.org/vol13/iss1/art25/
27. http://www.samveasna.org/
28. http://www.fairwild.org/documents/
29. http://www.trust.org/alertnet/blogs/asia-views/logging-concession-areas-good-for-orangutans-and-forest-conservation-says-study
30. http://www.orangutan.com/orangutan-news-new-study-says-logging-concessions-offer-sanctuary-for-orangutans/
31. http://www.itto.int/
32. https://www.cbd.int/sustainable/addis.shtml
33. Some of the State Forest Departments in India:
    1. Jammu and Kashmir: http://jkforest.com/jkforest/files/index.html
    2. Himachal Pradesh: http://hpforest.nic.in/
    3. Uttarakhand: http://uttarakhandforest.org/
    4. West Bengal: http://www.westbengalforest.gov.in/
    5. Sikkim: http://www.sikkimforest.gov.in/index.html
    6. Arunachal Pradesh: http://arunachalforests.gov.in/
    7. Manipur: http://manipurforest.gov.in/
    8. Assam: http://www.assamforest.in/common/
    9. Nagaland: http://nagaforest.nic.in/
    10. Mizoram: http://forest.mizoram.gov.in/
    11. Meghalaya: http://megforest.gov.in/
    12. Tripura: http://forest.tripura.gov.in/
    13. Tamilnadu: http://www.forests.tn.nic.in/WildBiodiversity/wildbiodiversity_home.html
    14. Maharashtra: http://www.mahaforest.nic.in/index.php
    15. Kerala: http://keralaforest.gov.in/
    16. Madhya Pradesh: http://www.mpforest.org/wildlife.html

34. *A Monograph of the Pheasants* by Dr. William Beebe.
    URL: http://www.biodiversitylibrary.org/bibliography/50777#/summary
35. Historical interest: http://thevictorianist.blogspot.in/2011/09/members-shall-discourage-wanton.html, http://www.19thc-artworldwide.org/index.php/spring02/206-qbeetle-abominationsq-and-birds-on-bonnets-zoological-fantasy-in-late-nineteenth-century-dress , http://www.academia.edu/1419871/Tales_Dead_Birds_Tell_the_historic

al_and_cultural_context_of_early_avian_specimens_in_the_biology_collecti
ons_of_Randolph_College

36. NORDECO: http://www.nordeco.dk/index.php?page=environmental-governance
37. Monitoring Matters: http://www.monitoringmatters.org/
38. Aarhus Convention, 1998:
    http://www.unece.org/env/pp/introduction.html
39. Aichi Biodiversity Targets: http://www.cbd.int/sp/targets/,
    http://www.cbd.int/2011-2020/goals/
40. Forest Rights Act, 2006: http://www.forestrightsact.com/,
    http://tribal.nic.in/Content/ForestRightActOtherLinks.aspx,
    http://www.fra.org.in/new/
41. Biological Diversity Act, 2002: http://www.nbaindia.org/
42. Peafowl Palaeontology: http://www.palass-pubs.org/palaeontology/pdf/Vol32/Pages%20439-446.pdf
43. American Museum of Natural History:
    http://diglib1.amnh.org/articles/chapin_bio/chapin_bio1.html
44. National Medicinal Plants Board: http://www.nmpb.nic.in/
45. Centre for Research, Planning and Action:
    http://www.cerpaindia.org/nmbp.php
46. *Heart of Darkness* by Joseph Conrad, 1899:
    https://ia600707.us.archive.org/27/items/HeartOfDarkness/conrad_jos
    eph_1857_1924_heart_of_darkness.pdf
47. *The Poisonwood Bible* by Barbara Kingsolver, 1998.
48. *My Life in Sarawak* by Lady Margaret Brooke, the Ranee of Sarawak,
    1913:
    http://www.archive.org/stream/cu31924021573468#page/n0/mode/2
    up
49. The Smithsonian's Museums of Asian Art – Freer Sackler Online
    Collection:
    http://www.asia.si.edu/collections/pinky.cfm?group=South Asian and
    Himalayan Art&start=570
50. Birding Tours: www.birdingindiatours.com/index.html,
    www.birdingtoursofindia.com/, www.indiabirdingtours.org/,
    www.indiabirding.com/, www.tragopantours.in/index.htm,
    indovacations.net/india-birding-tours/Index.htm,
    www.wildlifeindia.co.uk/index.html,
    www.birdtourasia.com/index.html

Addresses

1. CITES:

ADG (Wildlife) and CITES Management Authority,
Ministry of Environment and Forests (Wildlife Division)
Paryavaran Bhawan, CGO Complex
Lodhi Road, New Delhi - 110003
India
URL: http://www.cites.org/

2. IUCN, INDIA:

20 Anand Lok (2nd Floor)
August Kranti Marg
New Delhi - 110 049
 India
URL: http://iucn.org/

3. World Pheasant Association- India

WPA-India
782, Sector 17-A
Dwarka
New Delhi - 110078
Email: wpaindia@gmail.com
URL: http://www.pheasant.org.uk/

4. Wildlife Institute of India

Wildlife Institute of India
P.B.No. 18, Chandrabani,
Dehradun-248001
Uttrakhand, India
E-mail:- wii@wii.gov.in
URL: http://www.wii.gov.in

5. WWF - India

WWF - India
172 B, Lodhi Estate,
New Delhi- 110003
URL: http://www.wwfindia.org/

6. Wildlife Trust of India

Wildlife Trust of India
F-13, Sector-8, NOIDA,
Uttar Pradesh - 201301
India
URL: http://www.wti.org.in/oldsite/index.html

7. Central Zoo Authority, India

Central Zoo Authority
Annexe-VI, Bikaner House,
Shahjahan Road, New Delhi-110011
India
URL: http://cza.nic.in/

8. Wildlife Warden-cum-Deputy Conservator of Forests,
Wildlife Division, HP Forest Department,
Talland, Shimla - 171002
Himachal Pradesh, India
Telephone: 0177-2623993

Glossary

Adaptive management – The incorporation of a formal learning process into conservation action. Specifically, it is the integration of project design, management, and monitoring, to provide a framework to systematically test assumptions, promote learning, and supply timely information for management action.

Aviary – An enclosure constructed to house and rear birds. Unlike *cages*, aviaries allow birds a larger living space in which to fly and simulate a natural environment.

Aviculturalist – A person, who deals in aviculture, rears birds in aviaries as pets or for sale.

Biogeographic Realm – The world flora and fauna is divided into 6 biogeographic realms, their boundaries determined mainly by impassable barriers of climate and topography; each realm possesses distinctive forms of life unique to it. The 6 realms are – Nearctic, Palearctic, Oriental, Neotropical, Ethiopian and Australian realm. The realms were first proposed in 1858 by Philip L. Sclater and later modified in 1876 by Alfred Russel Wallace. The concept of realms also explains the historical biogeography of the earth.

Brood - A family of chicks that has hatched from one clutch of eggs.

Buttress – the flaring of the base of the trunk into folds seen in flood-plain trees like *Terminalia arjuna*, to give it extra stability.

Crop – A dialation of the oesophagus where food may be kept for a time before passing on to the gizzard or stomach. The term not only applies to birds but also to insects, earthworms and other invertebrates.

Clutch size – the number of eggs laid by one female in a nest at one time.

Congeneric – of the same genus.

Conspecific – of the same species.

Colonial – species occuring in a group of individuals during the breeding season (eg. shore birds) or throughout the lifecycle (eg. ants).

Dipterocarp – Mainly tropical lowland rainforest trees with two-winged fruits, belonging to the family Dipterocarpaceae. Many are large forest emergent species. Dipterocarpaceae is a family of 17 genera and approximately 580-680 species. The largest genera are *Shorea* (360 species), *Hopea* (105 species), *Dipterocarpus* (70 species), and *Vatica* (60 species). Their distribution is pantropical, from northern South America to Africa, the Seychelles, India, Indochina, and Malaysia, with the greatest diversity and abundance in western Malaysia. Some species are now endangered as a result of overcutting and extensive illegal logging.

Dust-bathing - is a comfort activity performed in many bird species, and in pheasants it may consists of several elements that may be present from the first day or appear after about two weeks after hatching. These include bill-raking, ground-scratching, wing-shaking (to sweep sand into the plumage), lying on the side, and head-rubbing. Dust bathing is frequently performed by individuals throughout life, especially if the proper warm, dry, and loose substrate materials are present.

Endemic – Indigenous; exclusively confined to a defined geographic area.

Ex situ – removed from its natural habitat.

Flagship species – popular charismatic species that serve as symbols to stimulate conservation awareness and action locally, nationally, regionally or globally.

Floristic region - Russian botanist Armen Takhtajan developed a system of floristic regions – geographical units of natural distribution - for flowering plants. He identified 6 floristic kingdoms containing thirty-five floristic regions, each of which is subdivided into floristic provinces, 152 in all.

Foraging – the act of searching for food and feeding by adult animals and birds.

Galliform – species belonging to the Order Galliformes.

Gizzard – A part of the alimentary canal having thick muscular walls often with hard horny teeth. Its main function is to break up hard food before it reaches the main digestive region. Gizzards are present in birds and in many invertebrates.

GPS – The Global Positioning System is a space-based global navigation satellite system that provides reliable location (latitude and longitude) and time information in all weather and at all times and anywhere on the Earth when and where there is an unobstructed line of sight to four or more GPS satellites. It is maintained by the United States government and is freely accessible by anyone with a GPS receiver.

Green Infrastructure – The green infrastructure of a city is made up of open spaces, which may be patches of natural forests, scrub, or wetlands that constitute the city's original landscape; or are man-made spaces such as, parks, ponds and river-side recreational areas. The conservation of urban green spaces is linked to sustainable environment in cities, both due to their impact on the physical environment and their impact on the attitudes of the city dwellers. Urban green spaces provide several 'ecological services' to the environment – such as, flood mitigation, temperature regulation, and control of dust and noise pollution. Urban green spaces also provide a home to the natural flora and fauna of the region, sometimes functioning as 'green corridors' connecting wildlife habitats. Not only do these spaces provide an opportunity for the public to experience and learn about the natural flora and fauna, and identify with the indigenous species, they also provide recreational spaces, thus improving the health of city dwellers. A green environment also helps to attract businesses, economic investment, and tourism, thus improving the economy of a city. Educational programmes such as 'guided nature walks', conducted on a regular basis in the cities, help to favourably change the people's lifestyle towards sustainability by influencing the choices they make on a daily basis. In Indian cities, the preservation of urban green infrastructure is the responsibility of the respective municipal governments. Some examples of green infrastructure are – Central Park and High Line in New York, USA, Boston Urban Wilds, USA, Supertrees in Singapore, Aravalli and Yamuna Biodiversity Parks in New Delhi, Green Belt forests in Shimla, Himachal Pradesh, and Cubbon Park in Bangalore, Karnataka, India.

Home range - Home range is a concept that can be traced back to a publication in 1943 by W.H. Burt, who constructed maps delineating the spatial extent or outside boundary of an animal's movement during the course of its everyday activities. The home range of an individual animal is typically constructed from a set of location points that have been collected over a period of time identifying the position in space of an individual at many points in time. Such data are now collected automatically using radio-collars placed on individuals.

Imprinting - a learning process in early life whereby species specific patterns of behavior are established.

Integrated Pest Management (IPM) means the careful consideration of all available pest control techniques and subsequent integration of appropriate measures that discourage the development of pest populations and keep pesticides and other interventions to levels that are economically justified and reduce or minimize risks to human health and the environment. IPM emphasizes the growth of a healthy crop with the least possible disruption to agro-ecosystems and encourages natural pest control mechanisms.

Invertebrates - Lower forms of life, or animals which do not have a vertebral column (backbone enclosing the spinal cord).

In situ – within the natural habitat in the historical range of a species.

IUCN Red List – a comprehensive inventory of the global conservation status of species, using a set of criteria to evaluate extinction risk (IUCN, 2001).

Lek – A display ground, usually referring to places where several gather to display.

Logging concession - Companies possessing logging concession licenses (that are granted by the government) within natural forests selectively choose and fell trees to try to ensure that the forests continue to grow and replace the trees that are harvested. It is different from forestry plantations, as deforestation (or open clearing) is prohibited in logging concession area, and timber is extracted in a sustainable manner. [In comparison, concessions for industrial oil palm plantations (IOPP) are granted by governments to allow the establishment and management of industrial, monoculture oil palm estates. An Industrial Tree plantation (ITP) concession is a right granted by a government to develop an area of land into an industrial monoculture timber plantation (e.g. *Acacia mangium, Hevea,* or *Eucalyptus* spp.).]

Megapodes - The megapodes, also known as incubator birds or mound-builders (because they lay eggs under a mound of earth letting the natural warmth of the sun or volcanoes incubate them), are stocky chicken-like birds with large feet, in the family Megapodiidae. Their name literally means *large foot* (Greek *mega* - large, *poda* - foot). Megapodes are superprecocial, hatching from their eggs in a very mature condition.

Miocene Epoch – A time period in the Neogene Period in Cenozoic Era in the geologic time scale, which lasted from about 7 million to about 23 million years ago, and saw the appearance of the first apes, and the evolutionary diversification of horses and mastodons.

Mitochondrial DNA - Mitochondrial DNA (mtDNA) is the DNA located in organelles called mitochondria, structures within eukaryotic cells that convert the chemical energy from food into a form that cells can use, ATP. Most other DNA present in eukaryotic organisms is found in the cell nucleus. The mtDNA is useful for studying the evolutionary relationships (phylogeny) of organisms. Biologists can compare mtDNA sequences among different species, and use the comparisons to build an evolutionary tree for the species examined.

Lure cropping - Growing crops like sorghum (fodder crop) as a 'lure' crop near the main crop such as pearl millet significantly reduce depredatory bird damage to the main crop. The results are further supported by food preference studies on rose-ringed parakeet in captivity.

Molting – Animals shed feathers, hair, or skin periodically, especially seasonally, in order to allow replacement of what is lost with new growth. The replacement of all or part of the feathers in birds is called a molt. Molts produce feathers that match the age and sex of the bird, and sometimes the season. Molting occurs in response to hormonal changes brought about by seasonal changes. The entire process is complex; a basic understanding of molting patterns can be a useful aid in identifying many species and in determining their age (The Cornell Lab of Ornithology, 2007).

Monitoring – regular, statistically-designed analysis of a population in order to record its numbers, composition, behaviour and distribution. Recurrent observations in a specific area designed to detect changes in numbers over time and hence determine whether a species is increasing or decreasing.

Monogamous – pairing and mating with one individual of the opposite sex.

Monograph - a detailed and documented treatise on a particular subject.

Monomorphic - having but a single form or appearance.

Montane – Montane is a biogeographic term which refers to highland areas located below the tree line. Montane regions generally have cooler temperatures and often have higher rainfall than the adjacent lowland regions, and are frequently home to distinct communities of plants and animals. Areas above the timberline are known as alpine regions.

Ocelli – 'Eye-spots' as found on the 'trains' of peafowl.

Ontogenesis – The whole sequence of development of an individual, from the fertilized ovum to mature adult.

Precocial - refers to species in which the young are relatively mature and mobile from the moment of birth or hatching. The opposite developmental strategy is called "altricial", where the young are born helpless. Precocial species are normally nidifugous, meaning that they leave the nest shortly after birth.

Polygamous - pairing with more than one female at a time.

Predator - An animal that attacks and feeds on other animals, normally kills several individuals during its life cycle.

Preening – is an act of spreading the secretions of the preen-gland and combing the feathers with the beak by birds. This helps in keeping the feathers waterproof and in good condition.

Re-introduction – an attempt to re-establish a species in an area that was once a part of their historical range, but from which they have become extinct (for any reason).

Roost – afternoon or night perch of a bird, usually on a tree branch, to rest and hide from predators.

Rotational set-aside land – Patches of farmland that are planted with insect-attracting plants so as to provide a suitable habitat for farmland galliform species during the breeding season.

Solitary – occurring singly.

Spur - a sharp horny projection on the legs of birds above the claws.

Stakeholders - are anyone who has an interest in the conservation/development project. Project stakeholders are individuals and organizations that are actively involved in the project, or whose interests may be affected as a result of project execution or project completion. They may also exert influence over the project's objectives and outcomes. The project management team must identify the stakeholders, determine their requirements and expectations, and, ensure their participation and support for a successful project. Typical examples of stakeholders include the government, NGOs, Sponsors, villagers (cattle-graziers, gatherers of medicinal plants/ bamboo, rattan etc/ firewood, honey-collectors), miners, builders, and tourism industry.

Surveys – extensive observations designed to establish the presence or absence of a species over a given area and its distribution in relation to physical factors such as altitude, slope, rainfall and biotic factors such as vegetation, predation.

Tarsus (plural tarsi) - part of the leg of a bird below the thigh.

Terrestrial – Living on the ground.

Transect - A transect is a path along which one records and counts occurrences of the object of study (e.g. animals/plants, noting each instance). It requires an observer to move along a fixed path and to count occurrences along the path and, at the same time, obtain the distance of the object from the path. This results in an estimate of the area covered, an estimate of the way in which detectability increases from probability 0 to 1 as one approaches the path. Using these two figures one can arrive at an estimate of the actual density of objects.

Trap cropping - Growing specific plants (eg. marigold *Tegetes* spp.) in strips of land in between crops, or on the circumference of fields, to kill or reduce soil nematodes, beetles, etc., that are parasitic to the crops. The 'trap plants' are toxic to the parasites.

Twitching - *Twitching* is a British term used to mean "the pursuit of a previously-located rare bird." In North America it is more often called "chasing", though the British usage is starting to catch on there, especially among younger birders. Twitching is highly developed in the United Kingdom, the Netherlands, Denmark, Ireland, Finland, and Sweden. The term *twitcher*, sometimes misapplied as a synonym for birder, is reserved for those who travel long distances to see a rare bird that would then be *ticked* or checked off on a list. The term originated in the 1950s, when it was used for the nervous behaviour of Howard Medhurst, a British birdwatcher. The main goal of twitching is often to accumulate species on one's lists. Some twitchers engage in competition to accumulate the longest species list. The act of the pursuit itself is referred to as a *twitch* or a *chase*. A rare bird that stays put long enough for people to see it is *twitchable* or *chaseable*. There are many other words in the terminology associated with twitching.

Uropygial gland – The 'preen-gland' on the dorsal side of a bird's tail, secreting oil with which the bird makes its feathers waterproof.

Wallace's Line – A line that separates the Oriental from the Australasian zoo-geographic region. It runs between the islands of Bali and Lombok, through the Macassar Strait between Borneo and Sulawesi, then south-east of the Philippine Islands. It is important because of the difference in the fauna found on either side. To the north-west of the line the mammals are all eutherians (higher mammals in which placenta occurs and the embryo reaches an advanced stage of development before birth) and include tigers, elephants, gibbons, orang-utans, bears, deer and rhino; birds include the peacock and the larger game birds. To the south-east of the line, including Australia, Tasmania and New Guinea, the characteristic mammals are the marsupials (mammals usually having a pouch in which they carry the young; a few are also found in America. The young are born in an extremely small and underdeveloped state and include kangaroos and wallabies), and monotremes (mammals which lay eggs - duck-billed platypus and echidna). No placentals are indigenous and the few that occur were probably introduced by man; birds include emus, cassowaries, lyre-birds and birds-of-paradise.

Wattle – Fleshy structure found on the face or around the eyes of birds.

World Heritage Site - The World Heritage List includes 962 properties forming part of the cultural and natural heritage which the World Heritage Committee considers as having outstanding universal value. These include 745 cultural, 188 natural and 29 mixed properties in 157 countries. As of September 2012, 190 States Parties have ratified the 1972 UN Convention concerning the Protection of the World Cultural and Natural Heritage.

List of Figures

Chapter 1

Fig. 1 Detail from a painting of western tragopan (*Tragopan melanocephalus*) male (on the rt.) and female, painted by Archibald Thorburn between 1918 and 1922, in William Beebe's 'A Monograph of the Pheasants' (created 1918-22), Dover Publications, Inc., NY.

Fig. 1.1 A sketch of the ring-necked pheasants by T.W. Wood, in 'Pheasants: Their Natural History and Practical Management' 2nd edition, 1881, by W.B. Tegetmeier. Source- Archive.org.

Fig 1.2 A map the 8 zoogeographical zones of the world, according to WWF . Source – Wikipedia: http://en.wikipedia.org/wiki/Biogeographical_realms

Fig. 1.3 A map showing the location of Democratic Republic of Congo in west Africa. Source – Wikipedia: http://en.wikipedia.org/wiki/Democratic_republic_of_congo

Fig 1.4 A photo of the Himalayan Mountains in the state of Himachal Pradesh in northern India, by Anita Chauhan, 2011. Source – Flickr.com: http://www.flickr.com/photos/30654468@N03/5717408637/in/set-72157626674715124

Fig. 1.5 Map showing the location of Indo-China region in Asia (Myanmar, Vietnam, Laos, Cambodia, Thailand, parts of China). Source – Wikipedia: http://en.wikipedia.org/wiki/Indo-China

Fig 1.6 A drawing of two-winged fruit, leaves, and flower of a Dipterocarpus sp. tree. Source – *Flora Javae* by C.L. Blume and J. B. Fischer, 1828, http://www.archive.org/details/mobot31753000659919

Fig. 1.7 Map showing the location of Malesia region in Asia. Source Wikipedia: http://en.wikipedia.org/wiki/Malesia

Fig. 1.8 Photo of the rainforest in Sumatra, Indonesia showing dense forest and emergent Dipterocarpus trees along a stream, by Stefan Ottomanski on Flickr.com: http://www.flickr.com/photos/stefansprograms/4677855674/in/faves-30654468@N03/

Fig. 1.9 Rhinoceros hornbill in tropical rainforest of Malaysia, by Kevin Lin, 2011, on Flickr.com: http://www.flickr.com/photos/hiyashi/5781217725/in/faves-30654468@N03/

Fig. 1.10 A sketch of Himalayan monal pheasants (female on the left, and males on the right), by T.W. Wood, in 'Pheasants: Their Natural History and Practical Management' 2nd edition, 1881, by W.B. Tegetmeier.

Fig. 1.11 Alpine forest and grassland habitat of the Himalayan monal, in the State of Uttarakhand in northern India. Photo by Bijoy Venugopal, 2009. Source: The Green Ogre - http://greenogreindia.org/encounter-the-himalayan-monal/

Fig. 1.12 Blood pheasant males and females in Khumbu, Nepal, by Peteris Erglis, 2006. Source -Flickr.com: http://www.flickr.com/photos/peteris_e/351706352/in/set-72157594469601299/

Fig. 1.13 A photo of koklass and kalij pheasant habitat in Water Catchment Area Sanctuary near Shimla, Himachal Pradessh, India, showing wet mixed temperate forest interspersed with streams and grasslands. By Anita Chauhan, 2010. Source – Flickr.com: http://www.flickr.com/photos/30654468@N03/4977007036/in/set-72157624862848512

Fig. 1.14 A sketch of great argus pheasants (female on the left, and males on the right), by T.W. Wood, in 'Pheasants: Their Natural History and Practical Management' 2nd edition, 1881, by W.B. Tegetmeier.

Fig. 1.15 A male kalij pheasant perched on a deodar tree in the Chhota Shimla locality of Shimla hills, Himachal Pradesh, India. Photo by Anita Chauhan, 2010. Source – Flickr.com: http://www.flickr.com/photos/30654468@N03/4950797711

Fig. 1.16 A male blue peafowl perched on a mango tree in the afternoon in New Delhi, India. Photo by Anita Chauhan, 2011. Source – Flickr.com: http://www.flickr.com/photos/30654468@N03/5517439250

Fig. 1.17 Blue peafowl males in the spotted deer enclosure in the National Zoological Park, by Anita Chauhan, 2009. Source – Flickr.com: http://www.flickr.com/photos/30654468@N03/4190508720

Fig 1.18 A. Cheer pheasant preening in Himalayan Bird Park, Shimla, H.P., India. B. Peahen preening, New Delhi, India. Photos by Anita Chauhan, 2010. Flickr.com: http://www.flickr.com/photos/30654468@N03/4976028779/in/photolist-8zHrZD-8zHrMt-dQhxKH-begoqa-cnEEy7 , http://www.flickr.com/photos/30654468@N03/4666499064/in/photolist-87n2Aq-arnQGa-85NLpZ-arnWwe-87KT5b-87KSTE-87KSuC-89WQzx-87iPQM-87iPBe-87n2cC-87GCTF-arnVUD-arqzd5-87KRvL-87nnHy

Fig. 1.19 Two male Himalayan monal pheasants in Uttarakhand, India. Photo by Bijoy Venugopal, 2009. Source: The Green Ogre - http://greenogreindia.org/encounter-the-himalayan-monal/

Fig. 1.20 Photo of a red junglefowl scratching amidst rhino dung-pile for beetles, Kaziranga National Park in Assam, India, March 2009, by Aditya Singh.

Fig. 1.21 A painting of a pair of Blyth's tragopan *Tragopan blythi* (male in the foreground) by Archibald Thorburn in 'A Monograph of the Pheasants' by William Beebe (created between 1918-1922), Dover Publications, Inc., NY. Source – Wikipedia: http://pl.wikipedia.org/wiki/Tragopan_%C5%BC%C3%B3%C5%82tolicy

Fig. 1.22 A. Rhododendron tree in bloom in Chhota Shimla, Shimla hills B. Berries of *Principia*, Shimla hills, Himachal Pradesh, India. Photos by Anita Chauhan. Source: http://www.flickr.com/photos/30654468@N03/sets/72157622382803429/with/57174 37093/

Fig. 1.23 A male Himalayan monal feeding under rhododendron bushes in Uttarakhand, India. Photo by Bijoy Venugopal, 2009. Source: The Green Ogre - http://greenogreindia.org/encounter-the-himalayan-monal/

Fig. 1.24 A walking trail in Shimla Water Catchment Area Sanctuary, Shimla, Himachal Pradesh, India. Photo by Anita Chauhan, 2010. Source: http://www.flickr.com/photos/30654468@N03/

Fig. 1.25 Kalij pheasant pair chorus together in the breeding season at morning or evening hours. One pair was seen calling at 5 PM while perched 6-7 m high on a deodar tree in Chhota Shimla, Shimla hills, Himachal Pradesh, India. Photo by Anita Chauhan, August 2010. Source: http://www.flickr.com/photos/30654468@N03/

Fig. 1.26 A diagram showing the winter and spring home ranges of a common pheasant in its open country habitat, by Anita Chauhan. Photo of common pheasant by Gary Noon, 2007. Source: http://en.wikipedia.org/wiki/Phasianus

Fig. 1.27 Svalbard rock ptarmigan affixed with a 20 g satellite tag on its back in Norway. Photo by Eva Fuglei, 2009. Source: http://www.galliformes-sg.org/gallnewspdf/G@llinformed5.pdf

Fig. 1.28 Photo of a red fox in the high altitude cold desert of Hemis National Park, in Trans Himalayan Jammu and Kashmir, India, March 2010, by Aditya Singh. Source: http://www.flickr.com/photos/8581486@N05/5561212943

Fig. 1.29 A sketch of a pair of Lady Amherst's pheasants by T.W. Wood in 'Pheasants: Their Natural History and Practical Management' 2nd edition, 1881, by W.B. Tegetmeier.

Fig. 1.30 A sketch of Reeve's pheasants by T.W. Wood in 'Pheasants: Their Natural History and Practical Management' 2nd edition, 1881, by W.B. Tegetmeier.

Fig. 1.31 A sketch of argus pheasant males displaying their plumage to females, by T.W. Wood in 'Pheasants: Their Natural History and Practical Management' 2nd edition, 1881, by W.B. Tegetmeier.

Fig. 1.32 A sketch of golden pheasants by T.W. Wood in 'Pheasants: Their Natural History and Practical Management' 2nd edition, 1881, by W.B. Tegetmeier.

Fig. 1.33 Blue peafowl male displaying to peahen in the dry summers of Ranthambhore National Park, June, 2010, Rajasthan, India. Photo by Aditya Singh.

Fig. 1.34 A. A painting of the 5 species of tragopans, showing distended gular lappets, by Henrik Gronvold in 'A Monograph of the Pheasants' (created 1918-22) by William Beebe. Source: http://en.wikipedia.org/wiki/Tragopan B. A male Satyr tragopan displaying by distending its gular lappet, and erecting its horns, in a breeding center of Galliformes in Braga, Portugal. Photo by Hugo Barbosa. Source: http://www.flickr.com/photos/26613698@N03/2496886502/

Fig. 1.35 A. A nest and B. Newly hatched chicks, of the ring-necked pheasant in Oregon, USA. Source: The China or Denny Pheasant in Oregon, by W.T. Shaw, 1908, http://www.archive.org/stream/chinaordennyphea00shawrich#page/n47/mode/2up

Fig. 1.36 A sketch of a pair of silver pheasants by T.W. Wood in 'Pheasants: Their Natural History and Practical Management' 2nd edition, 1881, by W.B. Tegetmeier.

Chapter 2 –

Fig. 2 A male golden pheasant from 'A Hand-book to the Game-birds' Vol. I and II, by W.R. Ogilvie-Grant, 1897. Source: http://archive.org/details/handbooktogamebiog02ogil

Fig 2.1 Photo of a male red junglefowl in Kaziranga National Park, Assam, India, in March 2009 by Aditya Singh.

Fig. 2.2 A. A brass peacock lamp from south India, photographed at the Tamil Nadu state emporium, C.P., New Delhi, by Anita Chauhan.

Fig. 2.3 A map showing the location of Micronesia, Melanesia, and Polynesia. Source: http://en.wikipedia.org/wiki/Micronesia

Fig. 2.4 A male blue peafowl perched on a mango tree, New Delhi, India. Photo by Anita Chauhan, 2010. Source: http://www.flickr.com/photos/30654468@N03/5517440206

Fig. 2.5 A. A male common/ring-necked pheasant. B. A male green/Japanese pheasant (the subspecies of common pheasant). Sources: http://en.wikipedia.org/wiki/Common_pheasant#Subspecies and http://en.wikipedia.org/wiki/Green_pheasant

Fig. 2.6 A.Painting titled – 'The Last of the Flush', and B. Painting titled – 'The Count' - from 'The Pheasant' in Fur and Feather Series, 1895, by H.A. Macpherson, AJ Stuart-Wortley, AI Shand. Illustrations by A. Thorburn. C. Painting titled – 'TOHO!' – from 'American Partridge and Pheasant Shooting', 1877, by Frank Schley.

Fig 2.7 A painting of pheasants in an aviary. Source: http://en.wikipedia.org/wiki/Reeves%27s_Pheasant

Fig. 2.8 An advertisement for an aviary manufacturer, from at the back of 'Pheasants: Their Natural History and Practical Management' 2nd edition, 1881, by WB Tegetmeier.

Fig. 2.9 Photos of A. the Natural History Museum, London, UK, B. the Indian Museum in Kolkata, India. Sources: http://en.wikipedia.org/wiki/Natural_History_Museum,_London and http://en.wikipedia.org/wiki/Indian_Museum

Fig. 2.10 A. Dr. T.C. Jerdon, and B. Mr. A.O. Hume. Portraits from "The Nests and Eggs of Indian Birds" Vol. 1, 1889, by A.O. Hume.

Fig. 2.11 Cover of the book 'A Monographs of the Pheasants', by William Beebe, created in 1918-22. Publ. by Dover Publications, Inc., NY.

Fig. 2.12 A painting of a pair of blood pheasants by Joseph Wolf from 'A monograph of the Phasianidae' by D G Elliot, 1872. Source: http://en.wikipedia.org/wiki/Blood_pheasant

Fig. 2.13 A book by Dr. P.A. Johnsgard titled 'The Pheasants of the World – Biology and Natural History', published in 1999 by the Smithsonian Institution Press, Washington, D.C.

Chapter 3

Fig. 3 Painting of a pair of blood pheasants *Ithaginis cruentus*, by H C Richter in 'A Century of Birds from the Himalayas' by John Gould (1830-32). Source: http://en.wikipedia.org/wiki/Blood_pheasant

Fig. 3.1 A. Blood pheasant female (brown), and male (grey, white and light green). Source: http://www.flickr.com/photos/peteris_e/1361224277/in/set-72157594469601299/ B. A male blood pheasant. http://www.flickr.com/photos/peteris_e/1362103640/in/set-72157594469601299/. Photos by Peteris Erglis, Nepal, 2006.

Fig 3.2 A. A male western tragopan, and B. A female western tragopan in Sarahan Pheasantry, Himachal Pradesh, India. Photos by Gobind Sagar Bhardwaj, 2006. Source: http://www.flickr.com/photos/13967067@N03/1618239036/ and http://www.flickr.com/photos/13967067@N03/1618239008/in/photostream/

Fig. 3.3 A male Satyr tragopan by Arjan Haverkamp, Artis zoo, Netherlands, 2007.   Source: http://www.flickr.com/photos/ajhaverkamp/417695593/

Fig. 3.4 A female Satyr tragopan in Central Park Zoo, New York. Photo by Staven, 2008. Source: http://www.flickr.com/photos/stavenn/2360310333/in/set-72157603574148386/

Fig. 3.5 A male Blyth's tragopan in San Diego Zoo, California. Photo by Ashok Khosla, 2007. Source: http://www.flickr.com/photos/khosla/402035070/

Fig. 3.6 A. A maleTemminck's tragopan in Columbus Zoo, Ohio, USA. Photo by Ray Anspach, 2008. Source: http://www.flickr.com/photos/17001485@N03/2853421647/in/set-72157607175707218/ B. A male Temminck's tragopan in a breeding center for

Galliformes in Braga, Portugal. Photo by Hugo Barbosa. Source:
http://www.flickr.com/people/26613698@N03/

Fig. 3.7 A. A male Cabot's tragopan in Bronx Zoo, New York, USA, 2008.
http://www.flickr.com/photos/stavenn/2691009464/in/set-72157603574148386/
B. A female Cabot's tragopan in Central Park Zoo, New York, USA, 2007.
http://www.flickr.com/photos/stavenn/941940950/in/set-72157603574148386/.
Photos by Staven.

Fig. 3.8 A. A male Himalayan monal in Central Park Zoo, New York, USA, 2008.
http://www.flickr.com/photos/stavenn/2332207931/in/set-72157603574148386/ B. A
female Himalayan monal in Central Park Zoo, New York, USA, 2008.
http://www.flickr.com/photos/stavenn/2289583257/in/set-72157603574148386/.
Photos by Staven.

Fig. 3.9 A male Chinese monal in Chengdu Wildlife Rescue Center, China, 2007. Photo by
Wu Jiawei. Source: http://www.flickr.com/photos/wujiawei/2119640199/in/set-
72157615074655431/

Fig. 3.10 A male Lafayette junglefowl in a breeding center for Galliformes in Braga,
Portugal. Photo by Hugo Barbosa. Source:
http://www.flickr.com/people/26613698@N03/

Fig. 3.11 A. A male black-crested kalij pheasant in Kaziranga National Park, Assam, India.
Photo by Aditya Singh, 2009. B. A male white-crested kalij pheasant feeding on a slope
covered with balsams in Chhota Shimla, Himachal Pradesh, India. Photo by Anita Chauhan.
Source: http://www.flickr.com/photos/30654468@N03/3979428735/in/set-
72157622388143541/

Fig. 3.12 A male silver pheasant in Portugal. Photo by Antonio Sardinha, 2007. Source:
http://www.flickr.com/photos/antonio_sardinha/2146038961/in/set-
72157601390023074/

Fig. 3.13 A male crestless fireback pheasant in the National Zoo of Malaysia. Photo by Li Xin,
May 2007.

Fig. 3.14 A. A close-up of the head of a male Bornean crested fireback pheasant, and B. A
male Bornean crested fireback pheasant, in a breeding center for Galliformes in Braga,
Potugal. Photos by Hugo Barbosa. Source:
http://www.flickr.com/people/26613698@N03/

Fig. 3.15 A male blue eared-pheasant in Central Park Zoo, New York, USA. Photo by Staven,
2007. Source: http://www.flickr.com/photos/stavenn/2069422845/in/set-
72157603574148386/

Fig. 3.16 A male Reeves's pheasant in an enclosure in China. Photo by Wu Jiawei.

Fig. 3.17 A close-up of a male golden pheasant in Columbus Zoo Ohio, USA. Photo by Ray
Anspach, 2008. Source:
http://www.flickr.com/photos/17001485@N03/2853423363/in/set-
72157607175707218/

Fig. 3.18 Close-up of a male Lady Amherst's pheasant showing ruffed neck feathers, in Bronx Zoo, New York, USA, 2008. Photo by Staven. Source: http://www.flickr.com/photos/stavenn/2881387557/in/set-72157603574148386/

Fig 3.19 A male mountain peacock pheasant in a breeding center for Galliformes, Braga, Portugal. Photo by Hugo Barbosa. Source: http://www.flickr.com/people/26613698@N03/

Fig. 3.20 A close-up of a male Palawan peacock pheasant in Bronx Zoo, New York, USA, 2008. Photo by Staven. Source: http://www.flickr.com/photos/stavenn/2685397601/in/set-72157603574148386/

Fig. 3.21 A male Germain's peacock pheasant , and B. A female Germain's peacock pheasant, in a breeding center for Galliformes in Braga, Portugal. Photos by Hugo Barbosa. Source: http://www.flickr.com/people/26613698@N03/

Fig. 3.22 A. A close-up of a male Malayan peacock pheasant in Bronx Zoo, and B. A male Malayan peacock pheasant in Bronx Zoo, New York, USA, 2007. Photo by Staven. Source for A. : http://www.flickr.com/photos/stavenn/420648286/in/set-72157603574148386/

Fig. 3.23 A close-up of a male Great argus pheasant, in the Bronx Zoo, New York, USA. Photo by Staven, 2008. Source: http://www.flickr.com/photos/stavenn/2532418077/in/set-72157603574148386/

Fig. 3.24 A close-up of a male green peafowl in Bronx Zoo, New York, USA. Photo by Staven, 2008. Source: http://www.flickr.com/photos/stavenn/2682636556/in/set-72157603574148386/

Fig. 3.25 A male Congo peafowl in the Bronx Zoo, New York, USA. Photo by Staven, 2007. Source: http://www.flickr.com/photos/stavenn/1199980432/in/set-72157603574148386/

Chapter 4

Fig. 4 A painting of a pair of koklass pheasants by H C Richter in '*The Birds of Asia*', vol. 7 by John Gould. Source: http://en.wikipedia.org/wiki/Koklass_Pheasant

Fig. 4.1 The 'Status Survey and Conservation Action Plan for Pheasants' by IUCN, 2000-2004. Source: http://data.iucn.org/dbtw-wpd/edocs/2000-075.pdf

Fig. 4.2 The 9 categories in the IUCN Red List. Source: http://www.iucnredlist.org/about/

Fig 4.3 Photo of oil palm tree with unripe and ripe fruit. Source: http://en.wikipedia.org/wiki/Oil_palm

Fig. 4.4 A painting of a male Bulwer's pheasant from 'A Hand-book to the Game-birds' vol. I and II, by W.R. Ogilvie-Grant, 1897. Source: http://archive.org/details/handbooktogamebi01ogil

Fig 4.5 Fans made from blue peafowl train feathers for sale on a pavement in New Delhi, India. Photo by Anita Chauhan, 2010. Source: http://www.flickr.com/photos/30654468@N03/4479049282/in/set-72157622556870123/

Fig 4.6 A Gaddi shepherd taking his herd of goats and sheep to alpine pastures, along the NH 22 in Himachal Pradesh. Photo by Anita Chauhan, April 2011. Source: http://www.flickr.com/photos/30654468@N03/5716631880/in/photolist-9HaeuC-9H7oZx-9HadL7-9HajbW-9HaitS-9Hmvga-ejeTvm

Fig 4.7 A. Fresh morel mushrooms, and medicinal herb *Viola canescens* in Kotgarh, Shimla district, Himachal Pradesh, B. Dried morel mushroom for sale in Lower Bazar, Shimla Mall, Himachal Pradesh, C. Dried morel mushroom for sale in Lower Bazar, Shimla Mall, Himachal Pradesh, India. Photos by Anita Chauhan, April, 2011. Source: http://www.flickr.com/photos/30654468@N03/5717148968 , http://www.flickr.com/photos/30654468@N03/5717302209 , http://www.flickr.com/photos/30654468@N03/5718378518

Fig 4.8 The FairWild label. Source: http://www.fairwild.org/certification-overview

Fig 4.9 A blue peafowl feather on the ground at the Yamuna Biodiversity Park, New Delhi. Photo by Anita Chauhan, December 2010.

Fig 4.10 A signboard at the entrance of the Himalayan Nature Park in Kufri, Himachal Pradesh, India. Photo by Anita Chauhan 2009. Source: http://www.flickr.com/photos/30654468@N03/3978171591

Fig 4.11 A male western tragopan in the circular pheasantry at Kufri Nature Park, Kufri, Himachal Pradesh, India. This male was brought from the pheasantry in Sarahan, H.P., where some success has been achieved in breeding this species using rescued specimens. Photo by Anita Chauhan, 2009. Source: http://www.flickr.com/photos/30654468@N03/

Fig 4.12 A. New conservation breeding enclosures at Khariun, Chail, B. Cheer pheasant at Khariun pheasantry. Chail, Himachal Pradesh, India. Photos by Satpal Dhiman.

Chapter 5

Fig. 5 A male green junglefowl from 'A Hand-book to the Game-birds' Vol. I and II, by W.R. Ogilvie-Grant, 1897. Source: http://archive.org/details/handbooktogamebiog02ogil

Fig 5.1 A male Cabot's tragopan from 'A Hand-book to the Game-birds' Vol. I and II, by W.R. Ogilvie-Grant, 1897. Source: http://archive.org/details/handbooktogamebi01ogil

Fig. 5.2 A. A male Satyr tragopan in a breeding center for Galliformes, in Braga, Portugal. Photo by Hugo Barbosa. Source: http://www.flickr.com/people/26613698@N03/ B. A female Satyr tragopan in Artis zoo, Netherlands, 2008. Photo by Arjan Haverkamp. Source: http://www.flickr.com/photos/ajhaverkamp/2358234684/

Fig 5.3 A male grey junglefowl from 'A Hand-book to the Game-birds' Vol. I and II, by W.R. Ogilvie-Grant, 1897. Source: http://archive.org/details/handbooktogamebiog02ogil

Fig 5.4 A male great argus pheasant from 'A Hand-book to the Game-birds' Vol. I and II, by W.R. Ogilvie-Grant, 1897. Source: http://archive.org/details/handbooktogamebiog02ogil

Fig 5.5 A male Reeves's pheasant from 'A Hand-book to the Game-birds' Vol. I and II, by W.R. Ogilvie-Grant, 1897. Source: http://archive.org/details/handbooktogamebiog02ogil

Chapter 6

Fig. 6 A painting of a pair of Mrs. Hume's pheasants from 'A Monograph of the Pheasants' by William Beebe, 1918-22. Publ. by Dover Publications, Inc., NY. Source: http://digitalgallery.nypl.org/nypldigital/dgkeysearchdetail.cfm?trg=1&strucID=114834&imageID=107354&total=202&num=160&word=11653&s=1&notword=&d=&c=&f=13&k=1&lWord=&lField=&sScope=Name&sLevel=&sLabel=Beebe%2C%20William&sort=&imgs=20&pos=166&e=w

Fig. 6.1 A map showing the States of India. Source: Office of the Registrar General and Census Commissioner, India website. URL: http://www.censusindia.gov.in/2011census/maps/administrative_maps/admmaps2011.html

Fig. 6.2 A map showing the 12 districts of the State of Himachal Pradesh. Source: Office of the Registrar General and Census Commissioner, India website. URL: http://www.censusindia.gov.in/maps/State_Maps/StateMaps_links/himachal01.html

Fig 6.3 A male koklass pheasant walking along the roadside in the Shimla Water Catchment Sanctuary, Himachal Pradesh, India. Photo by Anita Chauhan, 2010. Source: http://www.flickr.com/photos/30654468@N03/4950766455

Fig 6.4 The pheasantry at the Himalayan Nature Park in Kufri, Himachal Pradesh houses 8 species of pheasants. Photo by Satpal Dhiman.

Fig. 6.5 Map showing the location of Great Himalayan National Park, and the surrounding Protected Areas in Himachal Pradesh, India. Source: 'Livelihood changes in response to restrictions on resource extraction from the Great Himalayan National Park' by Priyanka Mohan Pisharoti (2008), MS dissertation submitted to Manipal University, Bangalore, India.

Fig 6.6 Photo of a pair of cheer pheasants in Uttarakhand, India by Kousik Nandy, 2012. Source: http://en.wikipedia.org/wiki/Cheer_Pheasant

Fig. 6.7 Map showing the distribution of cheer pheasants in India, By M. Veerappan. From: Sathyakumar, S. and Sivakumar, K. (Eds.). 2007. Galliformes of India. ENVIS Bulletin: Wildlife and Protected Areas, Vol. 10 (1). Wildlife Institute of India, Dehradun, India. 252.pp. URL: http://wiienvis.nic.in/PublicationDetails.aspx?SubLinkId=129&LinkId=627&Year=2007

Fig 6.8 A cheer pheasant family – male, female, and 2 chicks - in the older enclosures at Khariun, Chail. Photo by Satpal Dhiman.

Fig. 6.9 A graph showing the juvenile mortality data for cheer pheasants in Chail pheasantries, year 2005-2010, by Satpal Dhiman.

Fig 6.10 A. A view of the hexagonal pheasant enclosures at Khariun, Chail, Himachal Pradesh, B. Inside the new hexagonal enclosures at Khariun, Chail. Photos by Satpal Dhiman.

Fig 6.11 A veterinary doctor examining a cheer pheasant in the Khariun pheasantry, Chail, Himachal Pradesh. Photo by Satpal Dhiman.

Fig 6.12 A western tragopan male at the Himalayan Nature Park at Kufri, Himachal Pradesh, India. Photo by Anita Chauhan, 2009. Source: http://www.flickr.com/photos/30654468@N03/3978393953

6.13 A map showing the distribution of western tragopan in India, by M. Veerappan. From: Sathyakumar, S. and Sivakumar, K. (Eds.). 2007. Galliformes of India. ENVIS Bulletin: Wildlife and Protected Areas, Vol. 10 (1). Wildlife Institute of India, Dehradun, India. 252.pp. URL: http://wiienvis.nic.in/PublicationDetails.aspx?SubLinkId=129&LinkId=627&Year=2007

Fig 6.14 A. A female western tragopan, and B. a western tragopan chick at the Sarahan pheasantry, Shimla district, Himachal Pradesh, India. Photos by Satpal Dhiman.

Fig 6.15 A. The circular pheasantry pen and gazebo, and B. rectangular enclosures at Sarahan Pheasantry, Himachal Pradesh. Photos by Anita Chauhan, April 2011. Source: http://www.flickr.com/photos/30654468@N03/5718854580 , and http://www.flickr.com/photos/30654468@N03/5718265259/in/set-72157626674715124/

Fig. 6.16 A. Kashmir rock agama sunning, and B. Verditer flycatcher in Sarahan pheasantry, Himachal Pradesh. Photos by Anita Chauhan, 2011. Source: http://www.flickr.com/photos/30654468@N03/5718292305/in/set-72157626674715124/ and http://www.flickr.com/photos/30654468@N03/5718271297/in/set-72157626674715124/

Fig. 6.17 Bhima Kali Temple, Sarahan. Photo by Anita Chauhan, April 2011. Source: http://www.flickr.com/photos/30654468@N03/5718878082/in/set-72157626674715124/

Fig. 6.18 Painting of red junglefowl *Gallus gallus* pair from 'Indian Sporting Birds' by Frank Finn, 1915.

Fig. 6.19 A map showing the distribution of red junglefowl in India, by M. Veerappan. Source: From: Sathyakumar, S. and Sivakumar, K. (Eds.). 2007. Galliformes of India. ENVIS Bulletin: Wildlife and Protected Areas, Vol. 10 (1). Wildlife Institute of India, Dehradun, India. 252.pp. URL: http://wiienvis.nic.in/PublicationDetails.aspx?SubLinkId=129&LinkId=627&Year=2007

Fig 6.20 A koklass pheasant from 'A Hand-book to the Game-birds' Vol. I and II, by W.R. Ogilvie-Grant, 1897. Source: http://archive.org/details/handbooktogamebiog02ogil

Fig 6.21 A male blood pheasant from 'A Hand-book to the Game-birds' Vol. I and II, by W.R. Ogilvie-Grant, 1897. Source: http://archive.org/details/handbooktogamebiog02ogil

Fig 6.22 A male Sclater's monal by J G Keulemans in 'Proceedings of the Zoological Society of London', 1870. Source: http://en.wikipedia.org/wiki/Sclater%27s_monal

Fig 6.23 Painting of a pair of Mrs. Hume's pheasants from 'A Monograph of the Pheasants' by William Beebe, 1918-22. Publ. by Dover Publications, Inc., NY. Source: http://digitalgallery.nypl.org/nypldigital/dgkeysearchdetail.cfm?trg=1&strucID=114834&imageID=107354&total=202&num=160&word=11653&s=1&notword=&d=&c=&f=13&k=1&lWord=&lField=&sScope=Name&sLevel=&sLabel=Beebe%2C%20William&sort=&imgs=20&pos=166&e=r

Fig. 6.24 A map showing the distribution of grey junglefowl in India, by M. Veerappan. Source: From: Sathyakumar, S. and Sivakumar, K. (Eds.). 2007. Galliformes of India. ENVIS Bulletin: Wildlife and Protected Areas, Vol. 10 (1). Wildlife Institute of India, Dehradun, India. 252.pp. URL: http://wiienvis.nic.in/PublicationDetails.aspx?SubLinkId=129&LinkId=627&Year=2007

Appendix 7

Map showing the distribution of the pheasants from Dr. William Beebe's 'A Monograph of the Pheasants'. Publ. by Dover Publications, Inc., NY. Source: http://digitalgallery.nypl.org/nypldigital/dgkeysearchdetail.cfm?trg=1&strucID=114701&imageID=107220&word=beebe&s=1&notword=&d=&c=161,183,200,184,188,197,150&f=&k=1&lWord=&lField=&sScope=&sLevel=&sLabel=Nature%20%26%20Science&sort=&total=202&num=80&imgs=20&pNum=&pos=82

Front Cover

Clockwise from top – Himalayan monal pheasant by Stavenn, Golden pheasant by Ray Anspach, Lady Amherst's pheasant and argus pheasant by Staven NG, and a displaying blue peafowl and peahen by Aditya Singh. All photos are from Flickr.com and used with permission.

Back Cover

A pair of Reeves' pheasants from the book *Fancy Pheasants and their Allies* by Frank Finn, 1901. Source: Archive.org.

Appendix 1 – Red List Status

A list of pheasants that are globally threatened, from the IUCN Red List:

The conservation status of each group of birds and plants that appear on the Red List are assessed by groups of experts appointed by IUCN's Species Survival Commission. BirdLife International is the Red List Authority for all birds. BirdLife, WPA and the Galliform Specialist Group work together in compiling and analysing information on Galliformes, and assessing their Red List status. WPA staff and/or GSG, along with BirdLife staff are the official Red List evaluators for Galliformes. Further information can be found on the Red List website and the BirdLife International website.

Endangered

Edwards's pheasant	*Lophura edwardsi*
Vietnamese pheasant	*Lophura hatinhensis*
Bornean peacock-pheasant	*Polyplectron schleiermacheri*
Green peafowl	*Pavo muticus*

Vulnerable

Western tragopan	*Tragopan melanocephalus*
Blyth's tragopan	*Tragopan blythii*
Cabot's tragopan	*Tragopan caboti*
Sclater's monal	*Lophophorus sclateri*

Chinese monal	*Lophophorus lhuysii*
Hoogerwerf's (or Aceh) pheasant	*Lophura hoogerwerfi*
Salvadori's pheasant	*Lophura inornata*
Crestless fireback	*Lophura erythrophthalma*
Bulwer's (or wattled) pheasant	*Lophura bulweri*
Brown eared-pheasant	*Crossoptilon mantchuricum*
Cheer pheasant	*Catreus wallichi*
Reeves's pheasant	*Syrmaticus reevesii*
Mountain peacock-pheasant	*Polyplectron inopinatum*
Malaysian peacock-pheasant	*Polyplectron malacense*
Palawan peacock-pheasant	*Polyplectron emphanum*
Congo peafowl	*Afropavo congensis*

Near-threatened

Satyr tragopan	*Tragopan satyra*
Swinhoe's pheasant	*Lophura swinhoii*
Crested fireback	*Lophura ignita*
Siamese fireback	*Lophura diardl*
Tibetan eared-pheasant	*Crossoptilon harmani*
White eared-pheasant	*Crossoptilon crossoptilon*
Hume's pheasant	*Syrmaticus humiae*
Mikado pheasant	*Syrmaticus mikado*
Copper pheasant	*Syrmaticus soemmerringii*
Germain's peacock-pheasant	*Polyplectron germaini*
Crested argus	*Rheinardia ocellata*
Great argus	*Argusianus argus*
Elliot's pheasant	*Syrmaticus ellioti*

Appendix 2

Critically Endangered (CR)

A taxon is Critically Endangered when the best available evidence indicates that it meets any of the following criteria (A to E), and it is therefore considered to be facing an extremely high risk of extinction in the wild:

A.   Reduction in population size based on any of the following:

1.   An observed, estimated, inferred or suspected population size reduction of ≥90% over the last 10 years or three generations, whichever is the longer, where the causes of the reduction are clearly reversible AND understood AND ceased, based on (and specifying) any of the following:

(a)   Direct observation

(b)   An index of abundance appropriate to the taxon

(c)   A decline in area of occupancy, extent of occurrence and/or quality of habitat

(d)   Actual and potential levels of exploitation

(e)   The effects of introduced taxa, hybridization, pathogens, pollutants, competitors or parasites.

2.   An observed, estimated, inferred or suspected population size reduction of ≥80% over the last 10 years or three generations, whichever is the longer, where the reduction or its causes may not have ceased OR may not be understood OR may not be reversible, based on (and specifying)any of (a) to (e) under A1.

3.   A population size reduction of ≥80%, projected or suspected to be met within the next 10 years or three generations, whichever is the longer (up to a maximum of 100 years), based on (and specifying) any of (b) to (e) under A1.

4.   An observed, estimated, inferred, projected or suspected population size reduction of ≥80% over any 10 year or three generation period, whichever is longer (up to a maximum of 100 years in the future), where the time period must include both the past and the future, and where the reduction and its causes may not have ceased OR may not be understood OR may not be reversible, based on (and specifying) any of (a) to (e) under A1.

B. Geographic range in the form of either B1 (extent of occurrence) OR B2 (area of occupancy) OR both:

1. Extent of occurrence estimated to be less than 100 km², and estimates indicating at least two of a-c:

a. Severely fragmented or known to exist at only a single location.

b. Continuing decline, observed, inferred or projected, in any of the following:

(i) Extent of occurrence
(ii) Area of occupancy
(iii) Area, extent and/or quality of habitat
(iv) Number of locations or subpopulations
(v) Number of mature individuals

c. Extreme fluctuations in any of the following:

(i) Extent of occurrence
(ii) Area of occupancy
(iii) Number of locations or subpopulations
(iv) Number of mature individuals

2. Area of occupancy estimated to be less than 10 km², and estimates indicating at least two of a-c:

a. Severely fragmented or known to exist at only a single location.

b. Continuing decline, observed, inferred or projected, in any of the following:

(i) Extent of occurrence
(ii) Area of occupancy
(iii) Area, extent and/or quality of habitat
(iv) Number of locations or subpopulations
(v) Number of mature individuals

c. Extreme fluctuations in any of the following:

(i) Extent of occurrence
(ii) Area of occupancy
(iii) Number of locations or subpopulations
(iv) Number of mature individuals

C. Population size estimated to number fewer than 250 mature individuals and either:

1.   An estimated continuing decline of at least 25% within three years or one generation, whichever is longer, (up to a maximum of 100 years in the future) OR

2.   A continuing decline, observed, projected or inferred, in numbers of mature individuals AND at least one of the following (a-b):

a.   Population structure in the form of one of the following:

(i)   No subpopulation estimated to contain more than 50 mature individuals, OR

(ii)   At least 90% of mature individuals in one subpopulation.

b.   Extreme fluctuation in number of mature individuals.

D. Population size estimated to number fewer than 50 mature individuals.

E. Quantitative analysis showing that the probability of extinction in the wild is at least 50% within 10 years or three generations, whichever is the longer (up to a maximum of 100 years).

Thus, in the criteria for 'threatened' species, there is a hierarchical alphanumeric numbering system of criteria and sub criteria.

Examples of species and their Red List category and criteria -

Afropavo congensis - Vulnerable   C2a(i)

Lophura bulweri - Vulnerable   A2cd+3cd+4cd; C2a(i)

Lophura edwardsi - Endangered   B1ab(ii,iii,iv,v); C2a(i)

Catreus wallichi - Vulnerable   C2a(i)

Tragopan melanocephalus - Vulnerable   C2a(i)

Appendix 3*

A list of Indian States that have a pheasant species as their State Bird
State birds are awarded protection, are culturally important, and are a
symbol of the people's pride.

State	State Bird
1. Himachal Pradesh	Western Tragopan
2. Manipur	Hume's Pheasant
3. Mizoram	Hume's Pheasant
4. Nagaland	Blyth's Tragopan
5. Orissa	Blue Peafowl
6. Sikkim	Blood Pheasant
7. Uttarakhand	Himalayan Monal

Appendix 4*

Pheasants as National Birds

Country	National Bird
India	Blue Peafowl
Sri Lanka	Ceylon Junglefowl
Nepal	Himalayan Monal
Thailand	Siamese Fireback Pheasant
Burma	Green Peafowl (unofficial-Grey Peacock Pheasant)
Japan	Green Pheasant (subspecies of Common Pheasant)
France	Gallic Rooster (introduced)
Chile	Golden Pheasant (unofficial, introduced; official-Condor)

*Source: Wikipedia

Appendix 5

What you can do to help save the pheasants –

- ✓ Contribute to NGOs that are working at the community level in educating, and spreading awareness, providing alternative livelihood to people dependent on pheasant habitats, especially in the Himalayan states, and in other countries.

- ✓ Take membership of World Pheasant Association, and update yourself with the newsletter '*Mor*'. Contribute monetarily towards conservation projects.

- ✓ Use the internet to keep abreast with the latest in wildlife conservation. In your spare time, visit zoos, biodiversity parks, sanctuaries and national parks. Find out more about the species that you see, and their conservation status, on the internet. Be active in protecting your national wildlife resources. That is the best way you can help to save the species and their habitat.

- ✓ Buy books about wildlife and its conservation.

- ✓ Join Nature Clubs, Natural History Societies, and Bird-watching clubs. Help in spreading awareness about wildlife, and its conservation, by organizing programmes and competitions in schools and colleges.

- ✓ Buy, or grow vegetables and fruits treated with eco-friendly organic pesticides and fertilizers.

- ✓ Enroll for science courses at further education level, with subjects such as botany, zoology, taxonomy, ecology, and conservation science; and join conservation organizations, or State forest departments for a challenging career in the service of wildlife and nature.

- ✓ Make sure that the products that you buy are not made at the expense of natural forests. Buy only certified tea, coffee, cocoa, chocolate, herbal medicines, furniture, toys, musical instruments etc.

- ✓ Do not purchase products made from pheasant feathers and body parts. It is illegal.

Appendix 6

CITES (the Convention on International Trade in Endangered Species of Wild Fauna and Flora) is an international agreement between governments. Its aim is to ensure that international trade in specimens of wild animals and plants does not threaten their survival.

International wildlife trade is estimated to be worth billions of dollars per year, and to include hundreds of millions of plant and animal specimens. The trade is diverse, ranging from live animals and plants to a vast array of wildlife products derived from them, including food products, exotic leather goods, wooden musical instruments, timber, tourist curios, and medicinal plants. Levels of exploitation of some animal and plant species are high and the trade in them, together with other factors, such as habitat loss, is capable of heavily depleting their populations and even bringing some species close to extinction. Many wildlife species in trade are not endangered, but the existence of an agreement to ensure the sustainability of the trade is important in order to safeguard these resources for the future.

Because the trade in wild animals and plants crosses borders between countries, the effort to regulate it requires international cooperation to safeguard certain species from over-exploitation. CITES was conceived in the spirit of such cooperation. Today, it accords varying degrees of protection to more than 30,000 species of animals and plants, whether they are traded as live specimens, fur coats, or dried herbs.

CITES was drafted as a result of a resolution adopted in 1963 at a meeting of members of IUCN. The text of the Convention was finally agreed at in a meeting of representatives of 80 countries in Washington DC., USA, in March 1973; and in July 1975, CITES became operational.

CITES is an international agreement to which signatory countries adhere voluntarily. Countries that have agreed to be bound by the Convention ('joined' CITES) are known as Parties. Although CITES is legally binding on the Parties – in other words they have to implement the Convention – it does not take the place of national laws. Rather it provides a framework to be respected by each Party, which has to adopt its own domestic legislation to ensure that CITES is implemented at the national level.

For many years CITES has been among the conservation agreements with the largest membership, with now 178 Parties.

Appendices I, II and III to the Convention are lists of species afforded different levels or types of protection from over-exploitation. The IUCN Red List status of species is also consulted in grouping the animals and plants in the 3 CITES Appendices.

Appendix I lists species that are the most endangered among CITES-listed animals and plants (see Article II, paragraph 1 of the Convention). They are threatened with extinction and CITES prohibits international trade in specimens of these species except when the purpose of the import is not commercial (see Article III), for instance for scientific research. In these exceptional cases, trade may take place provided it is authorized by the granting of both an import permit and an export permit (or re-export certificate). Article VII of the Convention provides for a number of exemptions to this general prohibition.

Appendix II lists species that are not necessarily now threatened with extinction but that may become so unless trade is closely controlled. It also includes so-called "look-alike species", i.e. species of which the specimens in trade look like those of species listed for conservation reasons (see Article II, paragraph 2 of the Convention). International trade in specimens of Appendix-II species may be authorized by the granting of an export permit or re-export certificate. No import permit is necessary for these species under CITES (although a permit is needed in some countries that have taken stricter measures than CITES requires). Permits or certificates should only be granted if the relevant authorities are satisfied that certain conditions are met, above all that trade will not be detrimental to the survival of the species in the wild. (See Article IV of the Convention).

Appendix III is a list of species included at the request of a Party that already regulates trade in the species and that needs the cooperation of other countries to prevent unsustainable or illegal exploitation (see Article II, paragraph 3, of the Convention). International trade in specimens of species listed in this Appendix is allowed only on presentation of the appropriate permits or certificates. (See Article V of the Convention).

Pheasant species listed in CITES

Appendix 1

Catreus wallichii Cheer pheasant
Crossoptilon crossoptilon White eared pheasant
Lophophorus impejanus Himalayan monal/Impeyan monal
Lophophorus sclateri Sclaters monal
Syrmaticus humiae Hume's bar-backed pheasant
Tetraogallus tibetanus Tibetan snow cock
Tragopan melanocephalus Western tragopan
Tragopan blythii Blyth's tragopan

Appendix 2

Gallus sonneratii Grey jungle fowl
Ithaginis cruentus Blood pheasant
Polyplectron bicalcaratum Peacock pheasant

Appendix 3

Tragopan satyra Satyr tragopan

POLICY AND LAWS CONCERNING CITES IN INDIA

International trade in all wild fauna and flora in general, and the species covered under CITES in particular, is regulated jointly through the provisions of the Wildlife (Protection) Act 1972, the Foreign Trade (Development Regulation) Act 1992, the Foreign Trade Policy of Government of India and Customs Act, 1962. The Director of Wildlife Preservation, Government of India is the Management Authority for CITES in India.

Wildlife (Protection) Act, 1972

Hunting of wild animals has been prohibited under Sec. 9 of the Wildlife (P) Act, 1972. No person is allowed to hunt any wild animal specified in

Schedule I, II, III and IV except as provided under sections 11 and 12 of the Act. The Act also prohibits under section 17A, the collection or the trade in specified plants (whether alive or dead or part or derivative) i.e. those listed in Schedule VI of the Act, from any forest land and any area specified by notification by the Central Government. The Schedule VI of the Act lists all the six plants of Indian origin included in CITES appendices. Trade in Scheduled animals / animal article i.e. animals/animal articles covered under Schedule I and Part II of Schedule II (which also include some inverterbrate such as insects, corals, molluscs and sea cucumber) are prohibited under the said Act. Similarly, the Act disallows trade in all kinds of imported ivory, including that of African elephant. Export or import of wild animals and their parts and products is, however, allowed for the purpose of scientific research and exchange of animals between Zoos and is subject to licensing by the Director General of Foreign Trade (DGFT), Government of India. The Act has been amended in 2006 leading to the establishment of the National Tiger Conservation Authority and the Wild Life Crime Control Bureau (WCCB) with a statutory backing.

MANAGEMENT AUTHORITIES - CITES:

India is a signatory to CITES since 1976. The Additional Director General (Wildlife) cum Director, Wildlife Preservation, MoEF, Govt. of India is the Management Authority, CITES in India. There are 4 functional Assistant Management Authorities (AMA-CITES) working on behalf of CITES at Delhi, Calcutta, Chennai and Mumbai.

Scientific Authorities

Following are the Scientific Authorities to deal with the CITES related matter in the country.
1. Director, Zoological Survey of India, Kolkata (West Bengal)
2. Director, Botanical Survey of India, Kolkata (West Bengal)
3. Director, Central Marine Fisheries Research Institute, Cochin (Kerala)
4. Director, Wildlife Institute of India, Dehradun (Uttarakhand)

ENFORCEMENT

Considering the seriousness of organised Wildlife Crime having the inter-state and International ramification and illegal trade of Wildlife parts and products, the Wildlife Crime Control Bureau was created in 2007 under the provisions of the Wildlife Protection Act 1972. The Wildlife Crime Control Bureau, Head Quarter is situated at New Delhi and erstwhile Regional Wildlife Preservation offices have been designated as regional Wildlife Crime Control Bureau offices. These offices are located at New Delhi, Kolkata, Mumbai and Chennai. In addition there is a Wildlife Crime Control Bureau Office in Jabalpur, Madhya Pradesh. The Bureau is a multidisciplinary organisation with the following mandate given under 38 Z of the Wildlife Protection Act. Wildlife Crime Control Bureau is designated nodal agency for CITES related enforcement.

38 Z: Powers and functions of the Wildlife Crime Control Bureau-

(1) Subject to the provisions of this Act, the Wildlife Crime Control Bureau shall take measures with respect to –
(i) Collect and collate intelligence related to organized wildlife crime activities and to disseminate the same to State and other enforcement agencies for immediate action so as to apprehend the criminals and to establish a centralized wildlife crime databank;
(ii) Co-ordination of actions by various officers, State Governments and other authorities in connection with the enforcement of the provisions of this Act, either or through regional and border units set up by the Bureau;
(iii) Implementation of obligations under the various International Conventions and protocols that are in force at present or which may be ratified or acceded to by India in future;
(iv) Assistance to concerned authorities in foreign countries and concerned International organizations to facilitate co-ordination and universal action for wildlife crime control;
(v) Develop infrastructure and capacity building for scientific and professional investigation into wildlife crimes and assists State Government to ensure success in prosecutions related to wildlife crimes;
(vi) Advise the Government of India on issues relating to wildlife crimes and suggest changes required in relevant policies and laws from time to time;

(2) The Wildlife Crime Control Bureau shall exercise –
(i) Such powers as may be delegated to it and,

(ii) Such other powers as may be prescribed.

The enforcement of CITES provisions is presently being carried out by the Customs officials and Regional Deputy Directors, Wildlife Crime Control Bureau through the Customs Act, 1962 at the point of Import/Export and by the State Wildlife Departments headed by Chief Wildlife Wardens under the Wildlife (Protection) Act, 1972 elsewhere. The Chief Wildlife Wardens also issue licenses to the dealers and manufacturers of wildlife items and legal procurement certificate to the exporters of wild fauna and flora to establish the legality of the consignment in question. The Directorate of Revenue Intelligence (DRI), the Central Bureau of Investigation (CBI), the Police, the State Forest Departments and various other Central/State agencies also provide assistance in controlling illegal trade in wildlife.

Source: www.cites.org

Appendix 7

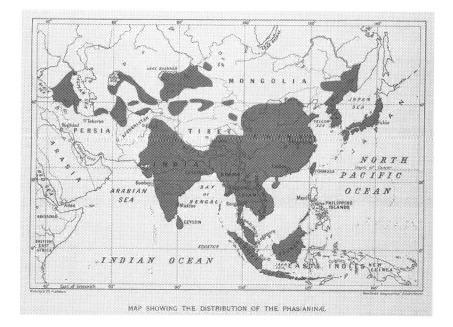

Map showing the extent of pheasant habitats in Asia, from Dr. William Beebe's 'A Monograph of the Pheasants' (1918-22).

Errata

The following figures should be attributed as -

Fig 1.2 A map the 8 zoogeographical zones of the world, according to WWF . Source – Wikipedia: Carol Spears, March 2008.
http://en.wikipedia.org/wiki/Biogeographical_realms

Fig. 2.5 B. A male green/Japanese pheasant (the subspecies of common pheasant) in Japan. Source: Coniferconifer, May 2009. http://en.wikipedia.org/wiki/Green_pheasant

Fig. 2.9 Photos of A. the Natural History Museum, London, UK, B. the Indian Museum in Kolkata, India. Sources: Valerie75, October 2007, http://en.wikipedia.org/wiki/Natural_History_Museum,_London and Michael Janich, June 2004, http://en.wikipedia.org/wiki/Indian_Museum

Fig. 2.10 A. Dr. T.C. Jerdon, and B. Mr. A.O. Hume. Portraits from 'The Nests and Eggs of Indian Birds' Vol. 1, 1889, by A.O. Hume. Source: Archive.org

Fig 4.3 Photo of oil palm tree with unripe and ripe fruit in a farm in Ajumaku, Ghana. Source: Bongoman, March 2008, http://en.wikipedia.org/wiki/Oil_palm

Fig. 6.18 Painting of red junglefowl *Gallus gallus* pair from 'Indian Sporting Birds' by Frank Finn, 1915. Source: Archive.org.

For all the sketches of pheasants by T.W. Wood in 'Pheasants: Their Natural History and Practical Management' 2nd edition, 1881, by W.B. Tegetmeier - Source: Archive.org